Linhares
**Development of Biodiesel-Resistant
Nitrile Rubber Compositions**

Felipe N. Linhares

Development of Biodiesel-Resistant Nitrile Rubber Compositions

HANSER

Hanser Publishers, Munich

Hanser Publications, Cincinnati

The Author:
Felipe N. Linhares, Ph.D., Polymer Processing Laboratory, Rio de Janeiro State University (UERJ)

Distributed in North and South America by:
Hanser Publications
6915 Valley Avenue, Cincinnati, Ohio 45244-3029, USA
Fax: (513) 527-8801
Phone: (513) 527-8977
www.hanserpublications.com

Distributed in all other countries by
Carl Hanser Verlag
Postfach 86 04 20, 81631 München, Germany
Fax: +49 (89) 98 48 09
www.hanser-fachbuch.de

The use of general descriptive names, trademarks, etc., in this publication, even if the former are not especially identified, is not to be taken as a sign that such names, as understood by the Trade Marks and Merchandise Marks Act, may accordingly be used freely by anyone. While the advice and information in this book are believed to be true and accurate at the date of going to press, neither the authors nor the editors nor the publisher can accept any legal responsibility for any errors or omissions that may be made. The publisher makes no warranty, express or implied, with respect to the material contained herein.

The final determination of the suitability of any information for the use contemplated for a given application remains the sole responsibility of the user.

Cataloging-in-Publication Data is on file with the Library of Congress

ISBN 978-1-56990-674-3
E-Book ISBN 978-1-56990-675-0

All rights reserved. No part of this book may be reproduced or transmitted in any form or by any means, electronic or mechanical, including photocopying or by any information storage and retrieval system, without permission in writing from the publisher.

© 2017 Carl Hanser Verlag, Munich
Coverdesign: Stephan Rönigk

DEDICATION

This Thesis is dedicated to my parents, who raised me, taught me the importance of education in life, and reminded me to continue studying even if I were tired.

ACKNOWLEDGMENTS

I thank my **supervisor**, **PROF. Dr. CRISTINA RUSSI GUIMARÃES FURTADO**, from **Universidade do Estado do Rio de Janeiro (UERJ)**, for her dedication, kindness, attention, friendship, trust, and for being a role model as a professor and as a researcher, not only while I conducted research for this Thesis, but also since I started as an undergraduate student in her laboratory. I wholeheartedly thank her for sharing parts of her knowledge with me, encouraging me throughout my graduate career, and having a direct influence on my training as a researcher. There are no words to describe the honour it gave me to be her student during these years, and the admiration I have for her.

I thank my **supervisor**, **PROF. Dr.–*Ing.* VOLKER ALTSTÄDT**, from **Universität Bayreuth**, for his trust, and for giving me the opportunity to conduct part of my doctoral research at the university.

I thank Prof. Marcia Amorim for her attention, trust, and friendship throughout these years, for her wise and comforting words, and for her willingness to always help everyone. Thank you for helping me to be a better person both professionally, and personally.

I thank Prof. Ana Furtado for her willingness to help me, answer my questions and to just have nice and relaxed conversations.

I thank Prof. Regina for her humility in sharing her knowledge, her kindness, and for her support on my academic path. One day, I hope I to have a fraction of her wisdom.

I thank Prof. Stefan Seifert for being part of this defence committee and for his help in improving the discussion on this Thesis topic.

I thank my sister, Juliana, for always being on my side, sometimes annoying me, sometimes advising me, but always supporting me.

I thank my dear sisters, Anna Christina and Mariana, for their kindness

I thank my large and beloved family, including my uncles, aunts, cousins, nephew, and godchildren, who are spread all over Brazil and the world, for teaching me what it is to love and to be loved. It would be impossible name each of you, and I would need more pages than this Thesis itself. Especially, I thank my beloved grandmothers (*in memoriam*).

I thank DJ, for thee kindness through these years.

I thank Fabiana and França for their kindness and for always being able to count on them.

I thank my "godchildren", Julyana, and Ostend, for simply being part of my life, for sharing one of the most important moments of their lives with me, and for putting up with me almost daily.

I thank my friends, Bruna, Susana, Diego, Carol, Luiz, for the support, for the friendship, for companionship and for sharing their joy with me. Without you all, my academic life would not have been worth it.

I thank Thomas for his friendship, for always supporting me, for the fun moments and for not complaining when I asked for help with English. Thank you for being a great friend, for your trust, and for making me admit that an American can make some of the best caipirinhas I have ever tried.

I thank my friends from the Laboratório de Processamento de Polímeros, Lessandra, Luciana, Leo, Cléverson, Thiago, Isaac, Monick, Haleck, Teresa and Gabi for making my afternoons at UERJ more fun, even when I would come back from the underground lab looking like a "coal boy".

I thank the friends from LEC/DQA, for the long hours that they had to put up with me while I was annoying them, especially Thais and Antônio

I thank the Brazucas in the Bay Area, Anna, Débora, Luiz, Pat, Daniel, Lucas, Bruna, Thiago, Isadora, Octávio, Gilvan, Thaís, Marcos, Paulo, and Ágatha, for having met them, and for making my staying in Bayreuth so comfortable as if I had never left Brazil.

I thank all my colleagues from the Polymer Engineering Department at the Universität Bayreuth, Theresa, Michaela, Edin, Gökhan, Amir, Ronak, Agustín, Stefanie, Stefan, Thomas, Daniel, Tobias, Rico, Bianca, Kerstin, Morris, Sebastian, Christian, Ute, Anne, Andreas, Markus, Jacqueline, Marko, Alexander, Robert, and Cathrin, thank you for the reception and for the kindness. I also hope you all liked the caipirinhas.

I thank everyone from DCI-UERJ, for their help, support, and kindness.

I thank all the professors from PPGQ, for given me the opportunity to conduct my doctoral research and for providing me with a valuable the learning experience.

I thank the PPGQ secretary, Luiz, for all his help.

I thank PPGEQ and its the professors, for their partnership, with special thanks to Prof. Lilian, for her wise words and kindness.

I thank IMA-UFRJ and especially all my colleagues from the Laboratório de Compostos de Borracha (Modulo 10) for their partnership, and help with the experimental analyses.

I thank the Federal Institute for Materials Research and Testing (BAM), especially Pedro Portella, Ute Niebergall, Martin Böhning and Marlon Zanotto, for the attention given to me, to their partnership, and for conducting the experimental analyses.

I thank UERJ and to IQ for being my second home, and sometimes my first home, during the 10 years that I am here.

I thank FAPERJ for the financial support.

I thank CAPES for the financial support.

I thank Nitriflex S/A Indústria e Comércio, and PETROBRAS for material supply.

I thank Universität Bayreuth for accepting me as an exchange student.

Ps 121:3

"Nobody said it was easy
No one ever said it would be this hard"
Guy Berryman, Jonny Buckland, Will Champion, Chris Martin (Coldplay)

"If education alone cannot transform society, without it society cannot change either"
Paulo Freire
"He will not let your foot slip He who watches over you will not slumber"
Ps 121:3

ABSTRACT

LINHARES, F. N. *Development of biodiesel-resistant nitrile rubber compositions.* 2016. 155f. Tese (Doutorado em Química) – Instituto de Química, Universidade do Estado do Rio de Janeiro, Rio de Janeiro; Fakultät für Angewandte Naturwissenschaften, Universität Bayreuth, Bayreuth, 2016.

Fossil fuels are the most commonly used sources of energy in the world. However, many reasons have pushed the society to search for alternative fuels to meet the world's energy demands without increasing environmental damage. Biodiesel has appeared to be an excellent replacement for petroleum diesel fuels because of its comparable physical properties in addition to its improved environmental benefits, such as low pollutant gas emissions, non-toxicity, renewability, and biodegradability. Nonetheless, biodiesel and petroleum diesel differ greatly with respect to their chemical properties. Therefore, the compatibility of the materials, which are commonly employed in contact with diesel, must also be assured for biodiesel that has been obtained from different sources. The main recommendation found in the literature is to withdraw nitrile rubber (NBR)-based articles from biodiesel applications. However, no effort has been made to better understand the interaction between nitrile rubber and biodiesel or to propose changes in and improvements to the production of nitrile rubber articles. This Thesis was devoted to evaluating the resistance of different types of NBR and different NBR formulations to biodiesel. First, nitrile rubbers with different acrylonitrile contents (33 and 45%), and carboxylated nitrile rubber (XNBR) with a 28% of acrylonitrile content were tested with soybean biodiesel. The preliminary results showed that an increase in acrylonitrile content increased the rubber resistance to biodiesel. Moreover, despite its low acrylonitrile content (28%), the carbolyxated nitrile rubber composition had almost the same performance as the NBR composition with high acrylonitrile content. This behaviour was probably due to the different types of crosslink network that XNBR was able to form during the vulcanisation process. Second, new formulations were prepared using high-acrylonitrile-content NBR by employing a two-level experimental design in which the amounts of two different accelerators (tetramethylthiuram disulphide-TMTD and N-cyclohexylbenzothiazole-2-sulphenamide - CBS), and the amount of sulphur were varied to achieve different types of vulcanisation systems (conventional, semi-efficient, and efficient). Statistically, the tensile strengths of the prepared compositions were influenced by TMTD, sulphur and the combination of TMTD and sulphur. Hardness was affected by the amount of TMTD, sulphur and the combination of CBS and sulphur, whereas elongation at break was affected by TMTD, sulphur, and the combination of both accelerators. The crosslink density was influenced only by the amounts of TMTD and sulphur. Furthermore, the choice of the accelerator played an important role on the resistance of nitrile rubber to biodiesel, and TMTD was found to be more effective than CBS with regards to the mechanical resistance to biodiesel. The crosslink density was not the only important factor with respect to the resistance. Nevertheless, dynamic mechanical thermal analyses showed that compositions prepared with an efficient vulcanisation system experienced chemical degradation of the crosslink network later than those prepared with conventional or semi-efficient vulcanisation systems. Based on these results, one could infer that nitrile rubber resistance can be still improved to meet the minimum requirements for these materials to be used in applications in which they will be in contact with biodiesel.

Keywords: Nitrile rubber. Biodiesel. Formulation. Compatibility. Curing system.

LIST OF ABBREVIATION

ANOVA	Analysis of variance
ANP	*Agência Nacional do Petróleo, Gás Natural e Biocombustíveis* (National Agency of Petroleum, Natural Gas and Biofuels)
ASTM	American Society for Testing and Materials
ATR	Attenuated total reflectance
CBS	N-cyclohexyl-2-benzothiazole sulphenamide
CI	Compression ignition
CLSM	Confocal laser scanning microscopy
CR	Polychloroprene rubber
CS	Carbon steel
CV	Conventional vulcanisation
DIN	*Deutsche Institut für Normung* (German Institute for Standardisation)
DMTA	Dynamic mechanical thermal analysis
DSC	Differential scanning calorimetry
EPDM	Ethylene propylene diene monomer rubber
EV	Efficient vulcanisation
FAAE	Fatty acid alkyl ester
FAEE	Fatty acid ethyl ester
FAME	Fatty acid methyl ester
FTIR	Fourrier transform infrared
GHG	Greenhouse gases
HDPE	High-density polyethylene
HNBR	Hydrogenated nitrile rubber
ISO	International Organization for Standardization
MDR	Moving die rheometer
LLDPE	Linear low density polyethylene
NBR	Acrylonitrile-butadiene rubber or Nitrile rubber
NMR	Nuclear magnetic resonance
OECD	Organisation for Economic Co-operation and Development
PE	Polyethylene
phr	Parts per a hundred of rubber

PTFE	Polytetrafluoroethylene
SEM	Scanning electron microscopy
SEV	Semi-efficient vulcanisation
SS	Stainless steel
TBBS	N-tert-butyl 2-benzothiazole sulfenamide
TMTD	Tetramethylthiuram disulphide
XLPE	Crosslinked polyethylene
XNBR	Carboxylated nitrile rubber

LIST OF SYMBOLS AND UNITS

α	Interaction parameter between the solvent and the elastomer
$°C$	Degree Celsius
C_H	Maximum torque
C_L	Minimum torque
C_t	Torque at a given time
E_a	Activation energy
g	Gram
J	Joule
G'	Elastic modulus or storage modulus
G''	Loss modulus or viscous modulus
k	Cure rate constant
K	Kelvin
m	Metre
M_1	Initial mass
M_2	Final mass
ΔM	Change in mass
min	Minute
n	Reaction order
N	Newton
R	Gas constant
T	Absolute temperature
t_{90}	Optimum cure time
t_{s1}	Scorch time
μ	Crosslink density
v_r	Volume fraction of rubber in equilibrium swollen vulcanizate sample
V_0	Molar volume
χ	Conversion

LIST OF FIGURES

Figure 1 – Biodiesel production per country in 2014 (in millions of m^3)....................22

Figure 2 – Biodiesel production via a fatty acid triglyceride transesterification reaction.28

Figure 3 – The molecular structure of nitirle rubber...38

Figure 4 – Sulphur network formation (vulcanisation)..40

Figure 5 – Simplified vulcanisation reaction scheme...42

Figure 6 – Generalised steps for vulcanisation..43

Figure 7 – Different types of crosslinks...43

Figure 8 – Interaction of Zn(II) with carboxyl groups from carboxylated nitrile rubber (XNBR)...45

Figure 9 – Chemical changes observed in nitrile rubber during thermal oxidation.........48

Figure 10 – Oxidation reactions of elastomers..49

Figure 11 – Summary of the Part I experimental section....................................64

Figure 12 – Summary of Part II experimental section.......................................65

Figure 13 – Crosslink densities of the nitrile rubber compositions vulcanised at 160°C......68

Figure 14 – Changes in mass of the nitrile rubber compositions after immersion in soybean biodiesel for 22 h at 100°C..69

Figure 15 – Changes in mass after 22 h of immersion in soybean biodiesel as a function of crosslink densities of the nitrile rubber compositions............................71

Figure 16 – Physical mechanical test results of non-immersed (dark colours) and immersed (light colours) nitrile rubber compositions. Between brackets indicate the percentage of loss of each composition...72

Figure 17 – Tensile strength change (%) after 22 h of immersion in soybean biodiesel at 100°C as a function of crosslink density of the nitrile rubber composition......74

Figure 18 – Tensile strength change (%) as a function of the mass uptake of each composition after 22h of immersion in soybean biodiesel at 100°C..............75

Figure 19 – Scanning electron microscopy (SEM) photomicrographs of nitrile rubber compositions: non-immersed and after immersion for 22 h at 100°C in soybean biodiesel...76

Figure 20 – Crosslink densities (bars) and glass transition temperatures (line) of the nitrile rubber compositions. Numbers between brackets indicate the amount in phr of each curing system component of the formulations (TMTD/CBS/sulphur).....79

Figure 21 – Mechanical test results of the nitrile rubber compositions. Between brackets indicate the amount in phr of each curing system component of the formulations (TMTD/CBS/sulphur)..81

Figure 22 – Tensile stregths (MPa) of the nitrile rubber compositions as a function of their crosslink densities. Numers between brackets indicate the amount in phr of each curing system component of the formulations (TMTD/CBS/sulphur)...........82

Figure 23 – Elongation at break and hardness as a function of crosslink density. Numbers between brackets indicate the amount in phr of each curing system component of the formulations (TMTD/CBS/sulphur)...83

Figure 24 – Elastic (a) and viscous (b) moduli *versus* temperature of the nitrile rubber compositions obtained at 1Hz. Numbers between brackets indicate the amount in phr of each curing system component of the formulations (TMTD/CBS/sulphur)..84

Figure 25 – SEM fracture surface photomicrographs of the nitrile rubber compositions. Numbers between brackets indicate the amount in phr of each curing system component of the formulations (TMTD/CBS/sulphur)...........................86

Figure 26 – SEM non-fractured surface photomicrographs of the nitrile rubber compositions. Numbers between brackets indicate the amount in phr of each curing system component of the formulations (TMTD/CBS/sulphur)...........87

Figure 27 – SEM fracture surface photomicrographs after cryogenic fracture of nitrile rubber composition with a high TMTD content. Composition *7(3/0/0.5)*.........88

Figure 28 – CLSM photomicrographs of the nitrile rubber compositions. Numbers between brackets indicate the amount in phr of each curing system component of the formulations (TMTD/CBS/sulphur)...89

Figure 29 – FTIR spectra of the nitrile rubber compositions with low and high TMTD content. Numbers between brackets indicate the amount in phr of each curing system component of the formulations (TMTD/CBS/sulphur)....................91

Figure 30 – Mass change profiles of the nitrile rubber compositions after 700 h of immersion in soybean biodiesel at 100°C. Numbers between brackets indicate the amount in phr of each curing system component of the formulations (TMTD/CBS/sulphur). In detail, the mass change profiles after the first 166 h...93

Figure 31 – Changes in mass after 22 h as a function of the crosslink densities of the nitrile rubber compositions. Numbers between brackets indicate the amount in phr of each curing system component of the formulations (TMTD/CBS/sulphur).....94

Figure 32 – Changes in mass after 700 h of immersion as a function of the crosslink densities of the nitrile rubber compositions. Numbers between brackets indicate the amount in phr of each curing system component of the formulations (TMTD/CBS/sulphur)..95

Figure 33 – Glass transition temperatures (°C) of the nitrile rubber compositions before and after 22 h, 46 h, 166 h, and 700 h of immersion in soybean biodiesel at 100°C. Numbers between brackets indicate the amount in phr of each curing system component of the formulations (TMTD/CBS/sulphur)..........................96

Figure 34 – Crosslink densities of the nitrile rubber compositions before and after heating ageing process for 22h at 100°C. Numbers between brackets indicate the amount in phr of each curing system component of the formulations (TMTD/CBS/sulphur)..97

Figure 35 – **(a)** Tensile strengths (MPa) of the nitrile rubber compositions: non-aged, after 22 h of ageing in air at 100°C, and after 22 h of immersion in soybean biodiesel at 100°C. **(b)** Relative change in tensile strengths of the nitrile rubber compositions after immersion in soybean biodiesel for 22 h at 100°C. Numbers between brackets indicate the amount in phr of each curing system component of the formulations (TMTD/CBS/sulphur)..98

Figure 36 – **(a)** Elongation at break (%) of the nitrile rubber compositions: non-aged, after 22 h of ageing in air at 100°C, and after 22 h of immersion in soybean biodiesel at 100°C. **(b)** Relative change in the elongation at break of the nitrile rubber compositions after immersion in soybean biodiesel for 22 h at 100°C. Numbers between brackets indicate the amount in phr of each curing system component of the formulations (TMTD/CBS/sulphur)..100

Figure 37 – Relative change in the tensile strength after immersion in soybean biodiesel for 22 h at 100°C as a function of the crosslink densities of the nitrile rubber compositions. Numbers between brackets indicate the amount in phr of each curing system component of the formulations (TMTD/CBS/sulphur).........101

Figure 38 - Relative changes in the tensile strengths after immersion in soybean biodiesel for 22 h at 100°C as a function of the vulcanisation systems of the nitrile rubber compositions. Numbers between brackets indicate the amount in phr of each curing system component of the formulations (TMTD/CBS/sulphur)..........102

Figure 39 – **(a)** Hardness (Shore A) of the nitrile rubber compositions: non-aged, (orange) and after 22 h of immersion in soybean biodiesel at 100°C (grey). **(b)** Relative changes in the hardness of the nitrile rubber compositions after immersion in soybean biodiesel for 22 h at 100°C. Numbers between brackets indicate the amount in phr of each curing system component of the formulations (TMTD/CBS/sulphur)……………………………………………………..104

Figure 40 – CLSM photomicrographs of the fracture surface of composition *3(1/2/0.5)*: (a) non-immersed and (b) after 22 h of immersion in soybean biodiesel at 100°C. Numbers between brackets indicate the amount in phr of each curing system component of the formulations (TMTD/CBS/sulphur)………………..105

Figure 41 – SEM photomicrographs of the fracture surface of composition *3(1/2/0.5)*: (a) non-immersed and (b) after 22 h of immersion in soybean biodiesel at 100°C. Numbers between brackets indicate the amount in phr of each curing system component of the formulations (TMTD/CBS/sulphur)………………..106

Figure 42 - SEM photomicrographs of the fracture surface of composition *5(3/0/0.5)*: (a) non-immersed, and (b) after 22 h of immersion in soybean biodiesel at 100°C. Numbers between brackets indicate the amount in phr of each curing system component of the formulations (TMTD/CBS/sulphur)………..……………..107

Figure 43 – SEM photomicrographs of the curing system's nitirle rubber compositions...126

Figure 44 – FTIR spectrum of pure TMTD……………………………………..127

Figure 45 – FTIR spectrum of pure CBS…………………………………………….128

Figure 46 – H^1NMR spectrum of the solubilised surface components of the nitrile rubber composition with low TMTD (1phr) in the formulation. Composition: 3 (1/2/0.5)…………………………………………………………………………129

Figure 47 – H^1NMR spectrum of the solubilised surface components of the nitrile rubber composition with high TMTD (3 phr) content in the formulation. Composition: 7 (3/2/0.5)…………………………………………………………………………130

Figure 48 – H^1NMR spectrum of the accelerator TMTD………………………..131

Figure 49 – H^1NMR spectrum of the accelerator CBS……………………………..131

Figure 50 –   Pareto chart of the stardardised effects variable on the mechanical properties. (a)Tensile Strength, (b)Elongation at break, and (c)Hardness....................133

Figure 51 –   Response surfaces for mechanical properties. (a) Tensile Strength; (b)Elongation at break; (c)Hardness...134

Figure 52 –   Pareto chart of the stardardised effects variable on the change in mass after 22h of immersion in soybean biodiesel at 100°C......................................135

LIST OF TABLES

Table 1 – Main biodiesel specifications according to Brazilian, European and USA regulations..30

Table 2 – Average weight percentage content of vegetable oils...............................30

Table 3 – Main accelerators used in sulphur vulcanisation of elastomers and their classifications...41

Table 4 – Different vulcanisation systems...44

Table 5 – Identification for each composition obtained from different nitrile rubber samples..54

Table 6 – Standard formulation recipe for nitrile rubber compositions according to ASTM D3187:11..55

Table 7 – Amount in phr[a] with respect to their coded amount...............................58

Table 8 – Design of experiments (2^3+3) for nitrile rubber with 45% of acrylonitrile content. . Between brackets indicate the amount in phr of each curing system component of the formulations (TMTD/CBS/sulphur)............................59

Table 9 – Main chemical and physical properties of soybean biodiesel (soybean methyl ester)...60

Table 10 – Rheometric parameters of the nitrile rubber compositions prepared with ASTM formulations (C_L – minimum torque; C_H – maximum torque; ΔC – maximum and minimum torques diferrence; t_{s1} – scorch time; t_{90} – optimum cure time)...66

Table 11 – Cure rate constant (k), reaction order (n), and activation energy (E_a) of the nitrile rubber compositions vulcanised at different temperatures.........................67

Table 12 – Vulcanisation systems of each prepared nitirle rubber composition. Numbers between brackets indicate the amount in phr of each curing system component of the formulations (TMTD/CBS/sulphur)..78

Table 13 – Glass transition tramperature (°C) of the compositions obtained from the tan maxima. Numbers between brackets indicate the amount in phr of each curing system component of the formulations (TMTD/CBS/sulphur)....................85

Table 14 – Glass transition temperatures (°C) of the nitrile rubber compositions before and after 22 h, 46 h, 166 h, and 700 h of immersion in soybean biodiesel at 100°C. Numbers between brackets indicate the amount in phr of each curing system component of the formulations (TMTD/CBS/sulphur)..........................96

Table 15 – Glass transition temperatures (°C) of the nitrile rubber compositions before and after 22 h, 46 h, 166 h, and 700 h of immersion in soybean biodiesel at 100°C. Numbers between brackets indicate the amount in phr of each curing system component of the formulations (TMTD/CBS/sulphur)..........................96

Table 16 - Chemical and physical properties of soybean biodiesel (Soybean methyl ester)...136

RESULTS OF THIS THESIS WERE PRESENTED IN THE FOLLOWING EVENTS

- 29th International Conference of the Polymer Processing Society (PPS29) – Nürnberg/Germany – July 15th–19th. *Resistance of Nitrile and Carboxylated-Nitrile rubbers to biodiesel.*
- 14th International Seminar on Elastomers – ISE 2014 – Bratislava/Slovakia – August 24th–28th. *Influence of accelerator content on nitrile rubber resistance to biodiesel.*

FULL PAPER PUBLISHED/ACCEPTED FOR PUBLICATION

- GABRIEL, C.F.S.; LINHARES, F.N.; SOUSA, A.M.F.; FURTADO, C.R.G.; PERES, A.C.C. *Vulcanization Kinetic Study of Different Nitrile Rubber (NBR) Compounds.* Macromolecular Symposia, v. 344, p. 22-27, 2014.
- LINHARES, F.N.; KERSCH, M.; SOUSA, A.M.F.; LEITE, M.C.A.M.; ALTSTÄDT, V.; FURTADO, C.R.G. *Influence of binary curing system on the nitrile rubber mechanical properties.* Macromolecular Symposia, v. 367, p. 55-62, 2016.
- LINHARES, F.N.; KERSCH, M.; NIEBERGALL, U.; LEITE, M.C.A.M.; ALTSTÄDT, V.; FURTADO, C.R.G. *Effect of different sulphur-based crosslink networks on the nitrile rubber resistance to bioiesel.* Fuel, to be published (Accepted in 2016).

PATENTS FILED

- BR 10 2015 05614 6. COMPOSIÇÕES ELASTOMÉRICAS RESISTENTES A BIODIESEL.

TABLE OF CONTENTS

	INTRODUCTION	22
1	LITERATURE REVIEW	25
1.1	**Biodiesel: an alternative to fossil fuels**	25
1.1.1	Overview on biodiesel	25
1.1.2	Biodiesel synthesis and possible feedstock	27
1.1.3	Physicochemical properties and oxidation stability	29
1.2	**Compression-ignition engines**	32
1.2.1	The compression-ignition engine and its composing materials	32
1.2.2	Compatibility of biodiesel with some compression-ignition engine parts	32
1.3	**Compatibility of biodiesel with elastomers**	33
1.4	**Nitrile rubber**	38
1.4.1	Main properties	38
1.4.2	Curing systems	39
1.4.3	Vulcanisation kinetics	45
1.4.4	Degradation process	47
2	AIMS	50
2.1	**General aims**	50
2.2	**Specific aims**	50
3	MATERIALS AND EQUIPMENT	51
3.1	**Part I – Preliminary studies: The influence of acrylonitrile content and different types of crosslink networks**	51
3.1.1	Materials	51
3.1.2	Equipment	51
3.2	**Part II – Formulation development: The influence of binary sulphur-based curing systems**	52
3.2.1	Materials	52
3.2.2	Equipment	52
4	METHODS	54
4.1	**Part I – Preliminary studies: The influence of acrylonitrile content and different types of crosslink networks**	54
4.1.1	Compounding	54
4.1.2	Vulcanisation	55

4.1.3	Vulcanisation kinetic	55
4.1.4	Crosslink density	55
4.1.5	Immersion tests	56
4.1.6	Change in mass	56
4.1.7	Mechanical tests	57
4.1.7.1	Strain-stress	57
4.1.7.2	Tear strength	57
4.1.7.3	Hardness	57
4.1.8	Scanning Electron Microscopy (SEM)	57
4.2	**Part II – Formulation development: The influence of binary sulphur-based curing systems**	**58**
4.2.1	Compounding	58
4.2.2	Vulcanisation	59
4.2.3	Crosslink density	59
4.2.4	Ageing tests	60
4.2.4.1	Ageing in air	60
4.2.4.2	Ageing in biodiesel	60
4.2.5	Gravimetric tests	60
4.2.6	Stress-strain	61
4.2.7	Hardness	61
4.2.8	Differential scanning calorimetry (DSC)	61
4.2.9	Dynamic mechanical thermal analysis (DMTA)	61
4.2.10	Scanning electron microscopy (SEM)	62
4.2.11	Confocal Laser Scanning Microscopy (CLSM)	62
4.2.12	Attenuated total reflectance Fourier transform infrared (ATR-FTIR) spectroscopy	62
4.2.13	Nuclear magnetic resonance (NMR)	62
4.2.14	Statistical analyses	63
4.3	**Experimental scheme**	**64**
4.3.1	Part I – Preliminary studies: The influence of acrylonitrile content and different types of crosslink networks	64
4.3.2	Part II – Formulation development: The influence of binary sulphur-based curing systems	65
5	**RESULTS AND DISCUSSION**	**66**

5.1	**Part I – Preliminary studies: The influence of acrylonitrile content and different types of crosslink networks**	**66**
5.1.1	Characterisation of the compositions	66
5.1.1.1	Vulcanisation kinetics	66
5.1.1.2	Crosslink density	68
5.1.2	Ageing tests	69
5.1.2.1	Gravimetric tests	69
5.1.2.2	Physical mechanical resistance	71
5.1.2.3	Scanning Electron Microscopy (SEM)	75
5.1.3	Overall performance	77
5.2	**Part II – Formulation development: The influence of binary sulphur-based curing systems**	**78**
5.2.1	Characterisation of the compositions	78
5.2.1.1	Crosslink density and differential scanning calorimetry (DSC)	78
5.2.1.2	Mechanical properties	80
5.2.1.3	Dynamic mechanical thermal analysis (DMTA)	84
5.2.1.4	Scanning electron microscopy (SEM)	85
5.2.1.5	Confocal Laser Scanning Microscopy (CLSM)	89
5.2.1.6	Attenuated total reflectance Fourier transform infrared (ATR-FTIR) spectroscopy	90
5.2.1.7	Nuclear magnetic resonance (NMR)	91
5.2.2	Ageing tests	92
5.2.2.1	Gravimetric tests	92
5.2.2.2	Differential scanning calorimetry (DSC)	95
5.2.2.3	Strain-stress	97
5.2.2.4	Hardness	103
5.2.2.5	Confocal Laser Scanning Microscopy (CLSM)	104
5.2.2.6	Scanning electron microscopy (SEM)	105
	CONCLUSIONS	**108**
	SUGGESTIONS FOR FUTURE WORKS	**111**
	REFERENCES	**112**
	APPENDIX A	**125**
	APPENDIX B	**126**
	APPENDIX C	**127**
	APPENDIX D	**129**

APPENDIX E..133

ANNEXE A 136

INTRODUCTION

Substantial efforts have been made to control pollutant gas emissions to minimise global warming effects. Based on the fact that these emissions are mainly a result of fossil fuels use, researchers are constantly motivated to find replacements for these pollutant fuels.

Biodiesel is currently the most ready-to-use substitute for petroleum diesel (PULLEN; SAEED, 2014; AKHLAGHI et al., 2015b), owing to its similar performance in compression ignition engines (ALI et al., 2016; MOSER, 2016; RASHED et al., 2016) to that of diesel and its lower level of pollutant gas emissions (SHAHIR; JAWAHAR; SURESH, 2015). Moreover, biodiesel is considered to be non-toxic, renewable, and biodegradable (MOFIJUR et al., 2013b; TORREGROSA et al., 2013; ÖZENER et al., 2014; DAUD et al., 2015).

The world's biodiesel production is rising quickly. According to the Organisation for Economic Co-operation and Development (OECD, 2015), the production of biodiesel has risen from 4×10^6 tons to 36×10^6 tons of biodiesel from 2005 to 2015. In 2014, the world's production (Figure 1) was led by the USA followed by Brazil, Germany and Indonesia (STATISTA, 2015).

Figure 1 – Biodiesel production per country in 2014 (in millions of m^3).

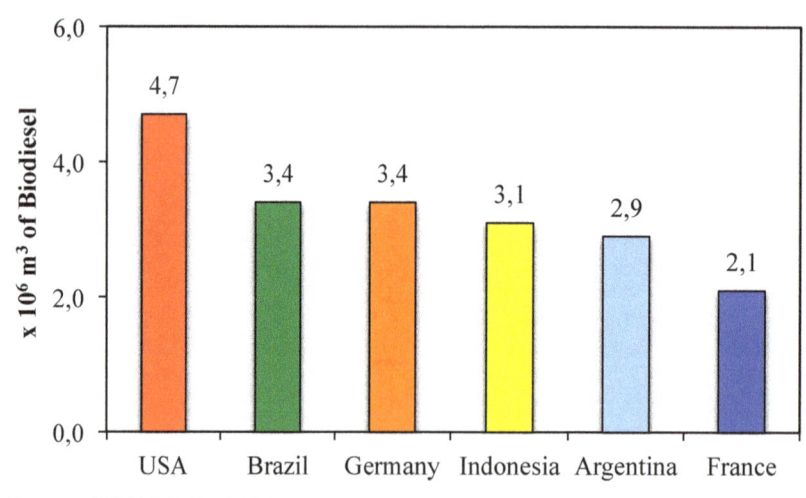

Source: STATISTA, 2015.

Biodiesel is no longer considered a "fuel of the future" but rather can be considered an actual fuel option. However, biodiesel is not commercially available in its pure form and is currently sold worldwide as a blend in small proportions with petroleum diesel. Blending biodiesel and petroleum diesel is mandatory in some countries: for example, Brazil implemented a policy in 2005 stating that diesel in petrol stations must be sold as a 2% biodiesel/diesel blend, and in 2014 the required concentration in the blend was raised to 7% (ANP, 2015). In the European Union, biodiesel consumption rose from 2.5 million tons of oil equivalent in 2005 to 11.5 million tons of oil equivalent in 2012 (STATISTA, 2015).

Nevertheless, despite the similarities and advantages between biodiesel and petroleum diesel, they possess significant chemical differences, which has raised concerns about material compatibility, especially with respect to their use in the automotive industry. Because of these differences, studies on the compatibility of biodiesel and several types of materials have become of vital importance (BÖHNING et al., 2014a; 2014b; CORONADO et al., 2014; FAZAL; HASEEB; MASJUKI, 2014; RESTREPO-FLÓREZ et al., 2014; AKHLAGHI et al., 2015b). To date, the compatibility of biodiesel with materials that are widely used in diesel engines has not been fully explained or understood.

Automotive units are built with a wide variety of materials, including metals and polymers (AKHLAGHI et al., 2015a; SORATE; BHALE; DHAOLAKIYA, 2015). Among the polymers used, there is also a wide diversity of plastics and rubbers that can be employed. Nitrile rubber (NBR) is widely employed in automotive parts that require constant contact with fuels and oils in automotive parts, e.g., hoses and seals, owing to its high polarity (AKHLAGHI et al., 2015a; AKHLAGHI et al., 2015b).

Nitrile rubber resistance has already been tested by several different authors under different conditions; a unanimous consensus has been made, stating that nitrile rubber presents low resistance to biodiesel (FRAME; BESSEE; MARBACH JR., 1997; FRAME; MCCORMICK, 2005; TERRY, 2005; TRAKARNPRUK; PORNTANGJITLIKIT, 2008; WALKER, 2009; ZHANG et al., 2009; HASEEB et al., 2010; CHAI et al., 2011; HASEEB et al., 2011b; ANDRIYANA et al., 2012; ALVES; MELLO; MEDEIROS, 2013; CHAI et al., 2013; LINHARES et al., 2013; CORONADO et al., 2014; AKHLAGHI et al., 2015b; CH'NG et al., 2015; DUBOVSKÝ; BOŽEK; OLŠOVSKÝ, 2015; ZHU et al., 2015). Nevertheless, most of the authors failed to provide complete information on the type of nitrile rubber that was tested, which impedes a thorough understanding of the interaction between the rubber and biodiesel. Fluorine-based elastomers are, however, highly resistant to different biodiesels and have consequently been proposed as a substitute for nitrile rubber in automotive parts. However, in Brazilian market, fluoro-elastomers are ten times more expensive than nitrile rubber (ALICE Web, 2015), and these higher prices of the material would have an impact on vehicles prices.

Furthermore, the authors did not make efforts to truly understand the interaction between biodiesel and nitrile rubber, or to improve the resistance of nitrile rubber to biodiesel, and thus avoiding changing the current use of cheaper raw materials for the production of most of automotive units.

The international appeal for knowledge regarding this topic resulted in the agreement that allowed this Thesis to be jointly supervised in Brazil, at the Universidade do Estado do Rio de Janeiro (UERJ), and in Germany, at the Universität Bayreuth.

The preliminary part of this Thesis discusses the resistance of different types of nitrile rubber to soybean biodiesel. The purpose of this part was to verify whether differences in the elastomer main structures would have an influence on the nitrile rubber resistance to biodiesel. All compositions were prepared using the same ASTM formulation, and the resistance of the rubber compositions was evaluated by gravimetric tests, and by measuring changes in the mechanical properties and morphologies of the compositions after immersing them in biodiesel. These preliminary tests were conducted to later decide which type of NBR sample would be chosen to conduct more thorough analyses of biodiesel resistance.

The first part of this Thesis was fully conducted in Brazil, at the Universidade do Estado do Rio de Janeiro (UERJ).

The second part of this Thesis contributes novel findings to this field. New formulations for nitrile rubber were prepared, and new approaches to evaluating the resistance of the prepared compositions were implemented. These new approaches included the assessment of the crosslink density and the effects of different vulcanisation systems on the resistance of nitrile rubber to biodiesel.

According to Haseeb et al. (2010; 2011a; 2011b), the biodiesel could interact with the elastomer curing systems; nonetheless, to date, there have been no publications that investigated the influence of a curing system on rubber resistance to biodiesel.

For this reason, different sulphur-based curing systems were employed, e.g. conventional, semi-efficient and efficient systems, using two different accelerators and sulphur in different amounts. The compounds were prepared and vulcanised in Brazil at the UERJ, and the material characterisations and immersion tests were conducted in Germany at the Universität Bayreuth.

1 LITERATURE REVIEW

1.1 Biodiesel: an alternative to fossil fuels

1.1.1 Overview on biodiesel

Fossil fuels are still the primary global source of energy, and their use is especially prominent in the transportation sector (CREMONEZ et al., 2015a; CREMONEZ et al., 2015b; MOHR et al., 2015; ULLAH et al., 2015), which contributes significantly to atmospheric pollution (SANGEETA et al., 2014). This sector alone is responsible for 15% to 26% of global greenhouse gas (GHG) emissions (ANDERSON, 2015; HOMBACH; WALTHER, 2015).

Different reasons have motivated the increasing interest in finding suitable replacements for fossil fuels over the past several decades. The oil crisis in the 70s, the Gulf war in 1991, recent oil price crises, the depletion of petroleum reserves, and concerns about the environment resulting from the high pollutant nature of fossil fuels are some of the factors that have encouraged the world to seek renewable and clean fuel sources (BERGMANN et al., 2013; MOFIJUR et al., 2013b; SAXENA; JAWALE; JOSHIPURA, 2013; ÖZENER et al., 2014; AVHAD; MARCHETTI, 2015; BERGTHORSON; THOMSON, 2015; CREMONEZ et al., 2015a; CREMONEZ et al., 2015b; SHAHIR; JAWAHAR; SURESH, 2015; YUNUS KHAN et al., 2015).

Considerable effort has been made to find a fuel that meets the world's present and future power demands without producing further environmental pollution (SANGEETA et al., 2014) or exhausting non-renewable natural resources (SAJID; KHAN; ZHANG, 2016). Different environmentally friendly sources of energy have been studied for use in fuel production; biodiesel appears to be a reasonable alternative to petroleum diesel because it produces lower GHG emissions compared to petroleum diesel and it can be easily manufactured (TORREGROSA et al., 2013; KAZAMIA; SMITH, 2014; SANGEETA et al., 2014; AVHAD; MARCHETTI, 2015; BERGTHORSON; THOMSON, 2015; HOMBACH; WALTHER, 2015; ULLAH et al., 2015; YUNUS KHAN et al., 2015; SAJID; KHAN; ZHANG, 2016).

G. Chavanne filed the first patent involving a chemical modification process to produce biodiesel in 1937 in Belgium. In 1977, the Brazilian scientist Expedito Parente filed the first patent on the industrial biodiesel process (MOFIJUR et al., 2013b; CREMONEZ et al., 2015b; DAUD et al., 2015).

Biodiesel is a renewable, biodegradable, non-toxic, and clean-burning fuel that is nearly carbon neutral, sulphur free, and aromatic free (MOFIJUR et al., 2013b; TORREGROSA et al., 2013; JAKERIA; FAZAL; HASEEB, 2014; ÖZENER et al., 2014; DAUD et al., 2015; SHAHIR; JAWAHAR; SURESH, 2015; ULLAH et al., 2015). In addition, it possesses physical properties similar to those of petroleum-diesel, e.g., viscosity, or even better, e.g., cetane number (MOFIJUR et al., 2013b), maintaining the same engine efficiency (SHAHIR; JAWAHAR; SURESH, 2015), and enhancing the lubrication of the engine (AGARWAL; CHHIBBER; BHATNAGAR, 2013; JAKERIA; FAZAL; HASEEB, 2014; NICOLAU et al., 2014).

The properties of biodiesel have been was thoroughly evaluated with respect to engine performance, and different exhaust pollutant emissions, e.g., NO_x, particulate matter (PM), carbon dioxide (CO_2), carbon monoxide (CO), unburnt hydrocarbons, and oxygenates (BERGTHORSON; THOMSON, 2015; CHONG et al., 2015). These emissions have been shown to be related to human health problems and environmental degradation (MOFIJUR et al., 2013a; ÖZENER et al., 2014).

It is widely agreed that CO_2 emissions are reduced when biodiesel is used instead of petroleum diesel (SHAHIR; JAWAHAR; SURESH, 2015). Nevertheless, Mattarelli, Rinaldini and Savioli (2015), and Özener et al. (2014) observed a slight increase in CO_2 emissions (approximately 5.5%). However, these increased emissions are balanced when the complete life-cycle of biodiesel is taken into consideration (GIAKOUMIS, 2013; ÖZENER et al., 2014). Reductions in the CO and PM emissions were usually observed (ANDERSON, 2015; BERGTHORSON; THOMSON, 2015; SHAHIR; JAWAHAR; SURESH, 2015), which were attributed to the increased presence of oxygen in biodiesel (SHAHIR; JAWAHAR; SURESH, 2015).

Özener et al. (2014) agreed with reports on reduced CO emission levels, but reported an increase in PM emissions. A 46% reduction of CO emissions was observed by Özener et al. (2014), when petroleum diesel was substituted by biodiesel produced from soybean oil. Moreover, Chong et al. (2015) reported findings on CO emission levels on an experiment that used petroleum diesel and biodiesel produced from palm oil.

The main disagreement lies on the inconclusive data obtained on NO_x emissions, which is one of the major concerns regarding the use of biodiesel as a fuel. Additionally, Chong et al. (2015) found comparable nitrogen monoxide (NO) emissions for petroleum diesel and palm oil biodiesel. Mattarelli, Rinaldini and Savioli (2015) observed no changes in NO_x emissions when petroleum diesel was replaced with biodiesel. Nevertheless, in general, an increase in emissions was noted, which was in agreement with findings reported by other authors (ÖZENER et al., 2014; ANDERSON, 2015; YUNUS KHAN et al., 2015). Özener, et al. (2014) reported an increase of up to approximately 18% in comparison to petroleum diesel emissions levels.

The observed divergences with respect to NO_x emissions cannot be attributed to a single reason (SHAHIR; JAWAHAR; SURESH, 2015). In addition to the higher presence of oxygen in biodiesel (ÖZENER et al., 2014), the operational settings of engines also influence the levels of NO_x emissions (CHONG et al., 2015); thus Shahir, Jawahar and Suresh (2015), Sangeeta et al. (2014), and Anderson (2015) stated that small changes in the engine settings obtain to optimum conditions could eliminate these problems.

According to Shahir, Jawahar and Suresh (2015), Özener et al. (2014), and Anderson (2015), biodiesel could reduce unburnt hydrocarbons emissions levels between 44%-55%, whereas oxygenated emissions, e.g. formaldehydes, were reported to increase (ANDERSON, 2015).

A major concerns in using biodiesel is the high risk of negative social impacts (EKENER-PETERSEN; HÖGLUND; FINNVEDEN, 2014). Its abundant production may compete with food production, leading to a possible food shortage in the long run (RIZWANUL FATTAH et al., 2014; CREMONEZ et al., 2015b; SHAHIR; JAWAHAR; SURESH, 2015; SAJID; KHAN; ZHANG, 2016); thus, non-edible resources have gained more attention for use as biodiesel feedstock (RIZWANUL FATTAH et al., 2014; YUNUS KHAN et al., 2014; MOFIJUR et al., 2015; SAJID; KHAN; ZHANG, 2016). Another crucial problem for biodiesel is its higher production cost compared to petroleum diesel (SAXENA; JAWALE; JOSHIPURA, 2013; CREMONEZ et al., 2015b; ULLAH et al., 2015).

Although biodiesel can be used in diesel engines in its pure form, with little to no modification (MOFIJUR et al., 2013b), it is more commonly available as a blend with petroleum diesel (ANDERSON, 2015), because they are fully miscible with each other (MOFIJUR et al., 2013b; JAKERIA; FAZAL; HASEEB, 2014). The blend is usually referred to as "BX", in which "X" indicates to the percentage of the biodiesel in the blend. Pure biodiesel is usually denoted as B100.

1.1.2 Biodiesel synthesis and possible feedstock

Biodiesel is chemically defined as a fatty acid methyl or ethyl ester (FAME, and FAEE, respectively), or, generally, as a fatty acid alkyl ester (FAAE), and it is obtained from the transesterification of vegetable oils, animal fats, or recycled waste oils (ARMAS; GÓMEZ; RAMOS, 2013; SAXENA; JAWALE; JOSHIPURA, 2013; SERRANO et al., 2013; RIZWANUL FATTAH et al., 2014). The name *biodiesel* should only be referred when a FAAE is used as an actual fuel in diesel engines (SERRANO et al., 2013)

Vegetable oils cannot be directly employed as a fuel in diesel engines because of their high viscosity, low volatility, and other unsuitable properties (MOFIJUR et al., 2013b; CREMONEZ et al., 2015b). Thus, they need to be refined to be turned into a valuable fuel (MOFIJUR et al., 2013b). Among the processes that are currently available, transesterification is the most reliable and commonly used process to overcome the problematic characteristics of vegetable oils (BERGMANN et al., 2013; MOFIJUR et al., 2013b; SANGEETA et al., 2014; CREMONEZ et al., 2015b; SAJID; KHAN; ZHANG, 2016).

Vegetable oils are composed of highly viscous fatty acid triglycerides, which are esters from glycerol and fatty acids. They can also contain mono- and di-glycerides and free fatty acids. The exact composition and content of vegetable oils varies according to the feedstock used (BERGMANN et al., 2013; CREMONEZ et al., 2015b).

Long chain fatty acid triglycerides react with short-chain alcohols, e.g., ethanol or methanol, in the presence of a catalyst (ULLAH et al., 2015). The FAAE obtained through this reaction is biodiesel, which is a product that is much less viscous and possesses properties similar to petroleum diesel. Glycerol is obtained as the by-product of the reaction (BERGMANN et al., 2013). The simplified reaction of vegetable oil to achieve biodiesel is shown in Figure 2.

Figure 2 – Biodiesel production via a fatty acid triglyceride transesterification reaction.

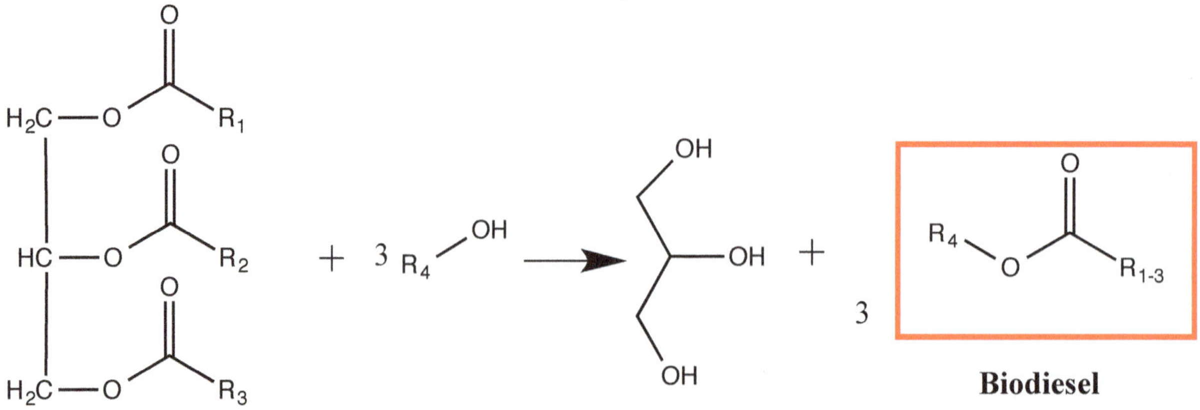

Footnote: R1, R2, R3 – represent long hydrocarbon chains (fatty acid triglyceride); R4 – represents short-chain hydrocarbon (methyl or ethyl).
Source: ULLAH et al., 2015 (adapted).

There are more than 350 different crops worldwide that can be used as possible sources for biodiesel production (MOFIJUR et al., 2013a). Vegetable sources that can be used for biodiesel production include soybean, palm, rapeseed, coconut, jatropha, karanja, sunflower, cottonseed, canola, castor bean, and babassu. (GIAKOUMIS, 2013; MOFIJUR et al., 2013b; SANTOS et al., 2013; AGARWAL et al., 2015; CHONG et al., 2015; CREMONEZ et al., 2015b; DAUD et al., 2015; MOFIJUR et al., 2015). Palm oil is the main feedstock used in some Asian countries, reaching 90% of the biodiesel production in Thailand (DAUD et al., 2015). Soybean oil comprises approximately 80% of Brazilian biodiesel production (SANTOS et al., 2013; ANP, 2015) and 75% of the production in the USA but only 16% of the European biodiesel production (SANTOS et al., 2013; SAXENA; JAWALE; JOSHIPURA, 2013), where the primary vegetable oil source used to produce biodiesel is rapeseed oil (BERGMANN et al., 2013).

The feedstock choice for biodiesel production varies for each country, and it is dependent on the local availability, climatic conditions, and economy (ULLAH et al., 2015). Avinash, Subramaniam and Murugesan (2014) listed the most used vegetable oil sources for biodiesel production in several countries around the world, as well as the economic and social impacts from biodiesel production in these countries.

1.1.3 Physicochemical properties and oxidation stability

Regardless of the environmental benefits attributed to biodiesel, it is necessary that it meet the minimum technical requirements to be accepted as ground transportation fuel (MOFIJUR et al., 2013b; CHONG et al., 2015). The American Society for Testing and Materials (ASTM) standard from the USA and the European standard are the two most referred standards worldwide, and other nations usually based their specifications on these standards (MOFIJUR et al., 2013b). Table 1 shows a comparison between the specifications of the USA, Europe and Brazil, where the specifications for biodiesel are regulated by the National Agency of Petroleum, Natural Gas and Biofuels (ANP).

Worldwide, soybean oil is the most commonly used feedstock for biodiesel production (GIAKOUMIS, 2013). Soybean oil, as well as the biodiesel obtained from it, is mainly composed of mono- and poly-unsaturated triglycerides, namely oleic acid, which is composed of 18 carbon-chain and 1 degree of unsaturation (C18:1), linoleic acid, which contains 18 carbons and 2 degrees of unsaturation (C18:2) and palmitic acid, which is a saturated acid

(C16:0) (GIAKOUMIS, 2013; SANTOS et al., 2013; SERRANO et al., 2013; RIZWANUL FATTAH et al., 2014). A significant amount of linolenic acid (C18:3) is also found in soybean oil. The fatty acids content of vegetable oils directly affects the physicochemical properties of the oils and consequently fuel quality (CREMONEZ et al., 2015b). Table 2 compares the triglycerides of soybean oil with two other important vegetable oil sources that are used worldwide for biodiesel production: rapeseed oil and palm oil.

Table 1 – Main biodiesel specifications according to Brazilian, European and USA regulations.

Property	Brazilian specification[I]	European specification[II]	USA specification[III]	Unit
Density (20°C)	850 – 900	860 – 900[a]	[b]	kg/m^3
Viscosity (40°C)	3.0 – 6.0	3.5 – 5.0	1.9 – 6.0	mm^2/s
Sulphur content	50 (max.)	10 (max.)	15 – 500[c] (max.)	mg/kg
Acid number	0.5 (max.)	0.5 (max.)	0.5 (max.)	mg KOH/g oil

Footnote: (a) – Density at 15°C; (b) – "The density of raw oils and fats is similar to biodiesel, therefore use of density as an expedient check of fuel quality may not be as useful for biodiesel as it is for petroleum based diesel fuel" (ASTM D6751:15, p. 6); (c) – Sulphur specification depends on the biodiesel (B100) grade used.
Source: (I) – ANP, 2015; (II) – DIN EN 14214:10; (III) – ASTM D 6751:15.

Table 2 – Average weight percentage content of vegetable oils.

Fatty acid	Chain length and degree of unsaturation	Soybean oil	Rapessed oil	Palm oil
Caprylic	C8:0	-	-	0.08
Capric	C10:0	-	-	0.06
Lauric	C12:0	0.08	-	0.37
Myristic	C14:0	0.12	0.04	1.13
Palmitic	C16:0	11.44	4.07	42.39
Palmitoleic	C16:1	0.16	0.23	0.17
Margaric	C17:0	-	0.07	0.06
Stearic	C18:0	4.14	1.55	4.20
Oleic	C18:1	23.47	62.24	40.91
Linoleic	C18:2	53.46	20.61	9.97
Linolenic	C18:3	6.64	8.72	0.29
Arachidic	C20:0	0.33	0.87	0.29
Gondoic	C20:1	0.22	1.09	0.16
Behenic	C22:0	0.27	0.27	-
Erucic	C22:1	0.07	0.71	-
Lignoceric	C24:0	0.13	-	0.05

Source: GIAKOUMIS, 2013 (adapted).

Another major technical concern for the use of biodiesel is its instability, i.e., low resistance to oxidation and polymerisation (GIAKOUMIS, 2013; SERRANO et al., 2013; JAKERIA; FAZAL; HASEEB, 2014; SANGEETA et al., 2014; BERGTHORSON; THOMSON, 2015), which can occur during storage in a tank system or during the combustion process (SANTOS et al., 2013; SERRANO et al., 2013; SANGEETA et al., 2014; SERQUEIRA et al., 2014; SINGER; RÜHE, 2014). This instability leads to altered fuel properties from the original ones, which can eventually result in a divergence from standard quality properties (JAKERIA; FAZAL; HASEEB, 2014; PULLEN; SAEED, 2014).

Biodiesel oxidation can be promoted in the presence of oxygen, water, or under other environmental conditions (SERQUEIRA et al., 2014; LANJEKAR; DESHMUKH, 2016). The degradation of biodiesel results in an increase in viscosity, coking of the injectors, damage to the fuel delivery system, filter plugging, engine deposits (gum-like sediments), automotive material corrosion, and other severe problems (PULLEN; SAEED, 2014; SANGEETA et al., 2014; SERQUEIRA et al., 2014; SORATE; BHALE, 2015). This oxidation process can be reduced by adding antioxidants or even by blending two or more types of biodiesel (SERQUEIRA et al., 2014).

The oxidation process occurs via radical chain reactions (SINGER; RÜHE, 2014) and can be divided into three phases (CHRISTENSEN; MCCORMICK, 2014) that are initiated by hydrogen abstraction (AKHLAGHI et al., 2015b), which leads to the formation of different types of alcohols, ketones, aldehydes, and carboxylic acids (JAKERIA; FAZAL; HASEEB, 2014; SANGEETA et al., 2014; SINGER; RÜHE, 2014). Further in-depth details regarding the degradation mechanism are given by Fattah et al. (2014), Singer and Rühe (2014), Jakeria, Fazal and Haseeb (2014), Christensen and McCormick (2014), Yaakob et al. (2014), and Akhlaghi et al. (2015b).

The degree of unsaturation is proportional to the oxidative stability of the fatty acids (SERQUEIRA et al., 2014), and the position and degree of conjugation of the double bonds also plays a role on the instability of the oil (SANTOS et al., 2013). Polyunsaturated fatty acids are more prone to oxidation than saturated fatty acids (GIAKOUMIS, 2013). Soybean oil is composed of 80% to 85% of unsaturated fatty acids, which reduces its oxidative stability (SANTOS et al., 2013). In contrast, coconut oil and palm oil are more stable, due to the high amounts of saturated fatty acids in their compositions (GIAKOUMIS, 2013).

Moreover, petroleum diesel is a fuel obtained from petroleum distillation and is difficult to characterise. The temperature of distillation ranges from 170°C to 350°C, and molecule sizes vary from 9 to 12 carbons (AGARWAL; CHHIBBER; BHATNAGAR, 2013).

Additionally, around 75% of its content consists of aliphatic hydrocarbons, and the remaining 25% consists of aromatic hydrocarbons (FAZAL; HASEEB; MASJUKI, 2014). Hence, petroleum diesel is less vulnerable to oxidative degradation than biodiesel (SERRANO et al., 2013; THOMPSON et al., 2013; RIZWANUL FATTAH et al., 2014).

1.2 Compression-ignition engines

1.2.1 The compression-ignition engine and its composing materials

Diesel engines, also known as compression-ignition (CI) engines, were named after their creator, Rudolf Diesel (BROWNSTEIN, 2015). The early engine could operate using pure vegetable oils, however, after the development and improvement of the engine over the years, this is no longer possible (ULLAH et al., 2015).

The high efficiency and reliability (TORREGROSA et al., 2013; UZUN, 2014; ZHANG; BALASUBRAMANIAN, 2014) of diesel engines made them widely preferred for many applications. Early petroleum diesel formulations had a high sulphur content, which made them unsuitable for use in passenger vehicles. However, changes implemented by the legislation of many countries required a reduction in the sulphur content of the fuel (TUDU; MURUGAN; PATEL, 2015), which permitted and spread the use of diesel engines in cars throughout the USA, Germany, and other European countries (BROWNSTEIN, 2015). Nonetheless, in some countries, such as Brazil, diesel engines are still only allowed in heavy-duty vehicles for transportation purpose.

Diesel engine systems consist of a variety of units, which come in contact with fuel: tank, fuel pump, fuel filter, injection nozzles, and fuel lines, among others (AKHLAGHI et al., 2015a; SORATE; BHALE, 2015). These units are made of a wide range of metallic and non-metallic materials, including aluminium (Al), copper (Cu), stainless steel (SS), carbon steel (CS), iron (Fe), synthetic rubbers, plastics, and others (CHEW et al., 2013; SORATE; BHALE, 2015).

1.2.2 Compatibility of biodiesel with some compression-ignition engine parts

Until now, most of studies conducted on biodiesel focused on its synthesis and the investigation of its properties rather than on its impact on typical materials that are used in the fuel systems of CI engines (THOMPSON et al., 2013; BÖHNING et al., 2014a).

Biodiesel can degrade easier and faster than petroleum diesel through oxidation during storage (SORATE; BHALE, 2013) because of its low stability and therefore is considered more corrosive than petroleum diesel (SORATE; BHALE, 2013; CURSARU et al., 2014). Because this corrosiveness can be drastic in a CI engine system (FAZAL; JAKERIA; HASEEB, 2014), a complete understanding of the interaction between biodiesel and various types of materials is of great importance (BÖHNING et al., 2014b; RESTREPO-FLÓREZ et al., 2014).

It is widely agreed that corrosion rates for metals are higher for biodiesel than for petroleum diesel; in addition, higher temperatures intensify corrosion attacks on metal surfaces (CHEW et al., 2013; CURSARU et al., 2014; JIN et al., 2015). Copper has proved to be less resistant to biodiesel compared to other metals and alloys, e.g., aluminium and carbon steel (CURSARU et al., 2014; FAZAL; JAKERIA; HASEEB, 2014). Chew et al. (2013) showed that magnesium exhibits poor resistance to palm biodiesel. The higher resistance of some metals or alloys can be explained by the formation of a passivation layer on the metal surfaces (KOVÁCS et al., 2015). Stainless steel, carbon steel and aluminium have been suggested as the most suitable metals for use in applications that come in contact with biodiesel (SORATE; BHALE, 2015). Moreover, Sorate and Bhale (2013), and Mojifur et al. (2013b) also reviewed the compatibility of other metals with different types of biodiesel.

Less attention has been given to the interaction between biodiesel and polyethylene (PE), which is one of the most commonly used polymer for the fabrication of fuel tanks (THOMPSON et al., 2013; BÖHNING et al., 2014b; RESTREPO-FLÓREZ et al., 2014). The few available studies tested different types of PE, e.g,. linear low density polyethylene (LLDPE), crosslinked polyethylene (XLPE), and high density polyethylene (HDPE). Petroleum diesel seemed to swell LLPE, XLPE (THOMPSON et al., 2013), and HDPE (BÖHNING et al., 2014a; 2014b) to a higher extent than biodiesel. Böhning et al. (2014a) mentioned that the swollen HDPE samples possessed higher molecular mobility and that this plasticisation effect changed the fracture mechanism and gave the material more ductile behaviour.

1.3 Compatibility of biodiesel with elastomers

The effects of biodiesel on different elastomers have been discussed over the years under different conditions. Although interesting conclusions have been drawn, the interaction between the two has not been fully explained.

Fluoroelastomers are considered to be fully compatible with biodiesel from different sources under any condition (FRAME; BESSEE; MARBACH JR., 1997; FRAME; MCCORMICK, 2005; TERRY, 2005; TRAKARNPRUK; PORNTANGJITLIKIT, 2008; WALKER, 2009; HASEEB et al., 2010; ALVES; MELLO; MEDEIROS, 2013). Mass and volume changes, after immersion in pure or blended biodiesel, were negligible regardless of the times lengths and temperatures of the tests. Trakarnpruk and Porntangjitlikit (2008) found an almost 4% change in volume after immersion in palm biodiesel for 1008 h at 60°C. the results of this observation were close to those reported by Terry (2005), who reported a variation of 6.8% in volume after immersing a sample in a B20 blend of soybean biodiesel for 1000 h at 60°C. Tensile strength was reportedly as unchanged (TERRY, 2005; HASEEB et al., 2010; ALVES; MELLO; MEDEIROS, 2013), or only slightly decreased (FRAME; BESSEE; MARBACH JR., 1997; TRAKARNPRUK; PORNTANGJITLIKIT, 2008) after immersion in biodiesel.

Similar comments can be extended to changes in other mechanical properties of fluoroelastomers. In addition, Walker (2009) showed that differences in fluorine content and curing systems affected the resistance to biodiesel. Nonetheless, the higher price of fluoroelastomers compared with other elastomers is a major concern and major obstacle for spreading their use in automotive parts (AKHLAGHI et al., 2015a).

Polychloroprene rubber (CR) showed poor resistance to biodiesel, especially in comparison with other rubbers (HASEEB et al., 2010; CHAI et al., 2011; HASEEB et al., 2011b; ANDRIYANA et al., 2012). Haseeb et. al (2010; 2011b) reported losses from 50% to over 80% after immersion in palm biodiesel. Although CR is no longer employed in parts that are in contact with fuels, it is used in hose covers, which may come in contact with fuels in case of leakage (AKHLAGHI et al., 2015a).

Ethylene propylene diene monomer rubber (EPDM) also showed little resistance to biodiesel, increasing almost 75% in volume, and losing approximately 80% of its initial tensile strength after immersion in palm biodiesel for 100 h at room temperature (HASEEB et al., 2011b).

The resistance of nitrile rubber has been exhaustively reported because of its importance in the automotive industry (FRAME; BESSEE; MARBACH JR., 1997; FRAME; MCCORMICK, 2005; TERRY, 2005; LINHARES; FURTADO, 2008; TRAKARNPRUK; PORNTANGJITLIKIT, 2008; WALKER, 2009; HASEEB et al., 2010; CHAI et al., 2011; HASEEB et al., 2011b; ANDRIYANA et al., 2012; ALVES; MELLO; MEDEIROS, 2013; LINHARES et al., 2013; CORONADO et al., 2014; AKHLAGHI et al., 2015b; CH'NG et al.,

2015; DUBOVSKÝ; BOŽEK; OLŠOVSKÝ, 2015; ZHU et al., 2015). Nonetheless most authors have failed to provide precise information on the formulation and grades of the NBR used (i.e. content of acrylonitrile), which makes it difficult to make a thorough comparison.

No significant changes to the mechanical properties of NBR were noted for samples with a high acrylonitrile content after immersion in low concentrations of biodiesel blends (i.e. up to B20 blends), even at different conditions and with different types of biodiesel (FRAME; MCCORMICK, 2005; TERRY, 2005; LINHARES; FURTADO, 2008; TRAKARNPRUK; PORNTANGJITLIKIT, 2008; WALKER, 2009; HASEEB et al., 2010). Frame and McCormick (2005), and Walker (2009) ran their tests at 40°C in soybean biodiesel B20 for 500 h, and for 1000 h, respectively. Both studies presented negligible changes in tensile strengths. Moreover, it is worth noting that the latter author found similar results for high acrylonitrile content NBR and for fluoroelastomers. Even at higher temperatures, the losses were moderate and were less than 20% (TERRY, 2005).

In contrast, Walker (2009), and Alves, Mello and Medeiros (2013) reported significant mechanical losses after immersion of samples in low concentration blends. However, the former study made it clear that these higher losses were noted for NBR samples with a 33% of acrylonitrile content. Linhares et al. (2013) and Walker (2009) showed that increasing the acrylonitrile content increased the resistance to biodiesel, regardless of the feedstock used.

To date, biodiesel is usually only available in the form of low concentrations blends, e.g., B5, B7 or B10, worldwide; thus, consumers can be confident in buying biodiesel/diesel blends from petrol stations without any concerns (LINHARES; FURTADO, 2008; TRAKARNPRUK; PORNTANGJITLIKIT, 2008).

When pure biodiesel (B100) was tested, NBR experienced higher degradation. Nonetheless some different results were observed, mostly likely because of differences in the tests conditions. Haseeb et al. (2010; 2011b) conducted tests at 25°C for 500 h and for 1000 h with palm biodiesel. They observed tensile strength losses of 15% and 22%, respectively. However, Frame, Bessee and Marbach Jr. (1997) found tensile strength losses of over 40% after immersion in soybean biodiesel at 50°C for 700 h. Nonetheless, no information was given on types of NBR samples that were used for these experiments.

Low-acrylonitrile-content NBR was shown to be weakly resistant to different types of biodiesel. Akhlaghi et al. (2015b) tested the resistance of NBR with 28% of acrylonitrile when immersed in rapeseed biodiesel at 90°C for 42 days, whereas Linhares at al. (2013) tested the same type of rubber in coconut biodiesel and castor bean biodiesel at 70°C for 70 h. The former reported a 40% biodiesel mass uptake and over 50% of elongation at break losses.

The latter observed tensile strength losses of approximately 80%. Nitrile rubber with 33% of acrylonitrile had between low (LINHARES et al., 2013) to moderate (WALKER, 2009) performance after immersion in biodiesel.

The degradation mechanisms of NBR by biodiesel have not been fully elucidated (HASEEB et al., 2011a), but it is known that biodiesel can act as a plasticiser and reduces chains entanglements (TRAKARNPRUK; PORNTANGJITLIKIT, 2008), hence decreasing the mechanical properties of the material. Although, many authors insist on using the "like dissolves like" principle as a key factor to explain the degradation of the rubber (HASEEB et al., 2010; HASEEB et al., 2011b; ALVES; MELLO; MEDEIROS, 2013; ZHU et al., 2015), the degradation process is not solely ruled by physical interactions, but rather by strong chemical interactions (CHAI et al., 2011).

Biodiesel's acid number tends to increase when it is subjected to heat (CORONADO et al., 2014; AKHLAGHI et al., 2015b) and indicates it has been oxidised. Carboxylic acids are formed as oxidation products, as described previously (JAKERIA; FAZAL; HASEEB, 2014; SANGEETA et al., 2014; SINGER; RÜHE, 2014). These carboxylic acids can chemically attack rubbers and promote its degradation (AKHLAGHI et al., 2015b). Furthermore, Zhu et al. (2015) showed that the molecular structure of the biodiesel, i.e., chain length and saturation, has a great influence on the swelling process of a rubber, when it is immersed in biodiesel. Increasing the carbon chain length decreases the swelling of NBR samples, which is in agreement with observations reported by Graham et al. (2006); additionally, NBR exhibits increased swelling in unsaturated fatty acid esters compared to that in saturated ones.

Haseeb et al. (2010; 2011b) speculated that biodiesel could also react with crosslink networks and filler systems of the rubber materials; nevertheless, no further tests were conducted to prove this hypothesis. Akhlaghi et al. (2015b) suggested that the crosslink densities of NBR samples undesirably increased when in contact with biodiesel. In addition, they also implied that biodiesel decreased the adhesion of carbon black particles on the rubber samples. Both findings negatively affect the mechanical properties of the rubber. This suggestion is in line with those reported by Mostafa et al. (2009), who concluded that carbon black decreased the thermal stability of NBR. In contrast, Ch'ng et al. (2015) suggested that fillers prevented biodiesel from diffusing into the rubber samples.

Increasing the crosslink density of rubber was suggested as a possible method to restrict the uptake of biodiesel by rubber (HASEEB et al., 2010; HASEEB et al., 2011b). Nevertheless, no tests were conducted to support this statement; this point will be discussed further in a later part of this Thesis. Little importance has been given to the different types of crosslink networks and different formulations for NBR, which leaves room for more investigations and opportunities to improve the properties of NBR-based automotive rubber parts.

Nitrile rubber resistance to biodiesel under special conditions was also investigated, e.g., under static mechanical strain (CH'NG et al., 2015), under cyclic loading (CHAI et al., 2011; ANDRIYANA et al., 2012; CHAI et al., 2013), and under pressure (ALVES; MELLO; MEDEIROS, 2013).

According to Akhlaghi et al. (2015a), hydrogenated nitrile rubber (HNBR) showed no improvement compared to regular NBR regarding its resistance to biodiesel, which is in line with observations reported by Walker (2009), but conflicts with those reported by Trakarnpruk and Porntangjitlikit (2008), and Terry (2005). Tests with acrylic rubber and silicone rubber (SR) have also been conducted and are reported elsewhere (TRAKARNPRUK; PORNTANGJITLIKIT, 2008; WALKER, 2009; HASEEB et al., 2011b; AKHLAGHI et al., 2015a).

Given the importance of rubber resistance to biodiesel, some patents have been filed over the years. An acrylate rubber composition was developed and claimed to have improved resistance to biodiesel. However, to achieve this improved resistance, the rubber composition required the inclusion of silicon rubber in its formulation (HONGQI; JIAN, 2014). Fluoroelastomers compositions patents, which includes organic peroxides as curing agents, were also filed, asserting their resistance to biodiesel and other hostile fluids (XU; RUI, 2012; LYONS; MORKEN, 2013).

Recently, a patent was filed related to an HNBR composition, which claimed to be biodiesel resistant (NASREDDINE; SODDEMANN, 2015) and whose formulation employed high levels of a filler. The invention was considered to have better resistance to aggressive fluids than other known HNBR compositions. Patents relating to compositions, whose formulations comprised nitrile rubber as the only elastomer, were not found. One patent related to a composition involving nitrile rubber was found, but fluoro-rubber was also present in the formulation (ZHANG; YANG, 2011).

1.4 Nitrile rubber

1.4.1 Main properties

Acrylonitrile-butadiene rubber, most commonly known as nitrile rubber (NBR), is a copolymer obtained by emulsion polymerisation (PAZUR; CORMIER, 2014) at high or low temperatures (OLIVEIRA et al., 2010). NBR is extensively employed in the automotive industry for manufacturing rubber-based goods, which requires constant contact with fuels (YASIN et al., 2003; ROCHA; SOARES; COUTINHO, 2007; ZHAO et al., 2013). The high resistance of NBR to mineral oils and non-polar solvents (FRANTA, 1989) is due to the presence of acrylonitrile (ACN) units in its structure (Figure 3).

The acrylonitrile content in nitrile rubbers usually varies from 18 – 50% and directly affects the properties of the elastomer. The properties affected include oil resistance, tensile strength, low and high temperature resistance, and hardness (FRANTA, 1989). In contrast to natural rubber (NR), NBR possesses no self-reinforcement effect because it does not crystallise under stress; thus it requires the use of fillers to achieve optimum mechanical properties (BARLOW, 1988; CIULLO; HEWITT, 1999; EL-NEMR, 2011).

Figure 3 – The molecular structure of nitirle rubber.

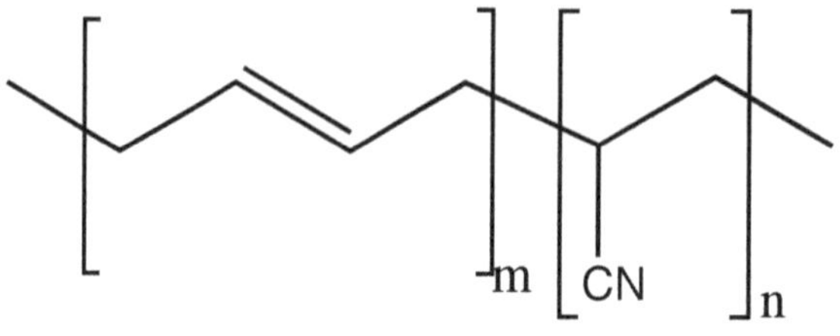

Modifications of the process for nitrile rubber syntheses are possible and are used to develop high-performance nitrile rubber for special applications (BHATTACHARJEE; BHOWMICK; AVASTH, 1991). Hydrogenated nitrile rubber (HNBR) is obtained by hydrogenating NBR (GUAN et al., 2011) and results in a material that has fewer free double bonds, which improves its resistance to heat and ageing (BHATTACHARJEE; BHOWMICK; AVASTH, 1991; CHOUDHURY; BHOWMICK; SODDEMANN, 2010). Carboxylated nitrile rubber (XNBR) can be obtained by adding a small amount of an unsaturated acrylic-type acid during the polymerisation process (FRANTA, 1989). The addition of carboxylic groups enhances some mechanical properties, e.g., hardness, and oil resistance (IBARRA; MARCOS-FERNÁNDEZ; ALZORRIZ, 2002).

1.4.2 Curing systems

Three major aspects should be considered in rubber compounding to allow the rubber articles to perform effectively during the required service period (CHANDRASEKARAN, 2010):

- The choice of the elastomer;
- The type and the level of reinforcement (filler);
- **The type of curing system that is employed to prepare the material.**

Curing is a key process in rubber technology (APREM; JOSEPH; THOMAS, 2005), and the sulphur curing process, better known as vulcanisation process, is the most popular method for rubber curing (NIEUWENHUIZEN, 2001; CORAN, 2003; CHOI; KIM, 2015). Other methods include the use of peroxides, metallic oxides, radiation, and others (APREM; JOSEPH; THOMAS, 2005; NIYOGI, 2007; IBARRA; RODRÍGUEZ; MORA-BARRANTES, 2008).

The curing, or vulcanisation, process can be defined as the formation of a three-dimensional elastic network (NIEUWENHUIZEN, 2001; IGNATZ-HOOVER; TO, 2004; CORAN, 2005), which chemically connects rubber chains, i.e., crosslinking (Figure 4). The curing process should possess a proper scorch time (delay) to allow shaping, forming and flowing of the material; afterwards, the crosslinking process should be rapid (CORAN, 2003). Vulcanisation is conducted in the presence of other chemicals, in addition to sulphur, i.e., accelerators and activators (APREM; JOSEPH; THOMAS, 2005; ANANDHAN et al., 2012).

The combination of zinc oxide (ZnO) and stearic acid is the most common and cost effective activator system employed in the sulphur vulcanisation process; this combination is used to enhance the effect of accelerators and thus reduces the curing time (FRANTA, 1989; DATTA; INGHAM, 2001; CORAN, 2003; COSTA et al., 2003; IGNATZ-HOOVER; TO, 2004; APREM; JOSEPH; THOMAS, 2005; SAHOO et al., 2007).

Accelerators are used in the rubber compounding process to significantly increase vulcanisation efficiency (FRANTA, 1989; DATTA; INGHAM, 2001; CORAN, 2003) and thus they play key roles in the vulcanisation technology (ALAM; MANDAL; DEBNATH, 2012b). A binary curing system is frequently employed to obtain a synergic, and more advantageous, effect (DEBNATH; BASU, 1996; COSTA et al., 2003; MARYKUTTY et al.,

2003; LAWANDY; HALIM, 2005; DATTA et al., 2007; ALAM; MANDAL; DEBNATH, 2012b; MOVAHED; ANSARIFAR; MIRZAIE, 2015). Mixed accelerator compounding can avoid pre-vulcanisation and allows the process to proceed faster (LAWANDY; HALIM, 2005).

Figure 4 – Sulphur network formation (vulcanisation).

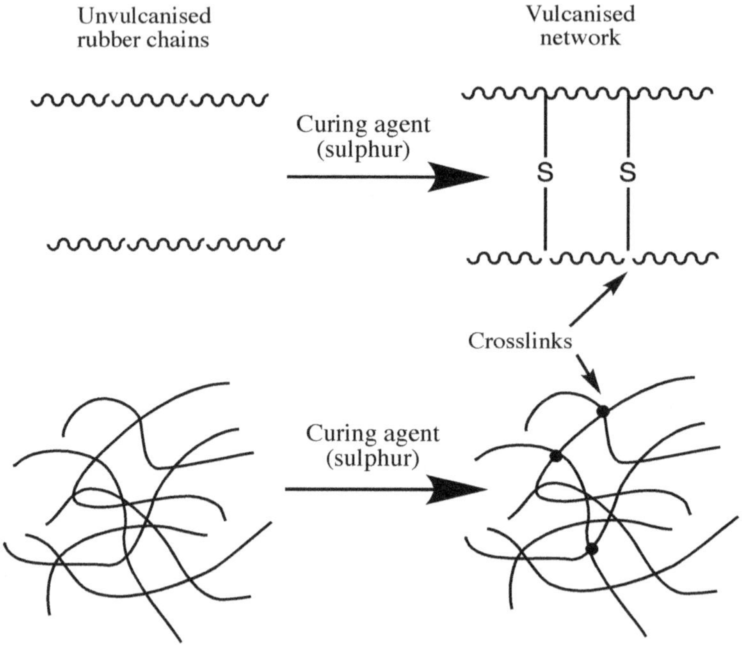

Source: CORAN, 2005 (adapted)

There are a wide variety of accelerators, which differ between depending on their effects on vulcanisation rates, scorching, ageing, and on the structure of the crosslinks formed (APREM; JOSEPH; THOMAS, 2005). Based on their chemical structures and actions, accelerators are classified into several different groups, and some are listed in Table 3. Table 15 in APPENDIX A shows a list of accelerators with their respective chemical structures.

The basic structures of most of accelerators are similar, which leads researchers to believe that the general steps in the vulcanisation reactions are applicable to a broad variety of accelerators (GHOSH et al., 2003; APREM; JOSEPH; THOMAS, 2005). Nevertheless, small changes in the chemical structures of the accelerators may have an important impact on their reactivities (APREM; JOSEPH; THOMAS, 2005).

Table 3 – Main accelerators used in sulphur vulcanisation of elastomers and their classifications.

Generic chemical designation	Property	Chemical name	Abbreviation
Thiazole	Fast	2-mercaptobenzothiazole	MBT
		2-2'-dithiobenzothiazole	MBTS
Sulphenamide	Fast – delayed action	N-t-butyl benzothiazole-2-sulphenamide	TBBS
		N-cyclohexyl benzothiazole-2-sulphenamide	CBS
Thiuram	Very fast	Tetramethyl thiuram monosulphide	TMTM
		Tetramethyl thiuram disulphide	TMTD
Dithiocarbamate	Ultra-accelerators	Zinc dimethyldithiocarbamate	ZDMC
		zinc diethyldithiocarbamate	ZDEC

Source: DATTA; INGHAM, 2001; COSTA et al., 2003; GHOSH et al., 2003; APREM; JOSEPH; THOMAS, 2005; CORAN, 2005; ALAM; MANDAL; DEBNATH, 2012.

Sulphenamide accelerators, e.g., TBBS (N-t-butyl benzothiazole-2-sulphenamide) and CBS (N-cyclohexyl benzothiazole-2-sulphenamide), are known for having delayed action (DEBNATH; BASU, 1996; GHOSH et al., 2003; APREM; JOSEPH; THOMAS, 2005; NIYOGI, 2007; ANANDHAN et al., 2012; MOVAHED; ANSARIFAR; MIRZAIE, 2015). Usually, they are employed in binary systems, especially with thiuram accelerators (MOVAHED; ANSARIFAR; MIRZAIE, 2015). Thiurams, as TMTD (Tetramethyl thiuram disulphide), can improve reversion resistance, especially in binary systems (DATTA et al., 2007). In addition, TMTD contains 13.31% available sulphur (HOFMANN, 1989). However, the use of CBS was found to decrease crosslink densities, even in binary vulcanisations systems (DEBNATH; BASU, 1996). Moreover, TMTD can bloom on the surface of rubber products, when used at high concentration (FRANTA, 1989).

Until now, there has not been a solid agreement on the correct mechanism for the sulphur-accelerated vulcanisation process (NIEUWENHUIZEN et al., 1999; APREM;

JOSEPH; THOMAS, 2005; MARZOCCA; MANSILLA, 2007; DONDI et al., 2015). Additionally, the synergetic activity of binary vulcanisation has yet to be fully elucidated. It is believed that the formation of new chemical moieties occurs during the reaction (SUSAMMA; MINI; KURIAKOSE, 2001; MARYKUTTY et al., 2003; ALAM; MANDAL; DEBNATH, 2012b).

Ghosh et al. (2003) divided the general vulcanisation mechanisms into three sub-categories: *(i) accelerator chemistry; (ii) crosslinking chemistry; and (iii) post-crosslinking chemistry*. This general view of vulcanisation is shown in Figure 5 and Figure 6. The first step involves the formation of an activator-accelerator complex, which then reacts with sulphur to generate monomeric polysulphide species, which are the actual active sulphurating agents. These agents finally react with rubber and form crosslink bonds (CORAN, 2003; GHOSH et al., 2003; APREM; JOSEPH; THOMAS, 2005; ANANDHAN et al., 2012).

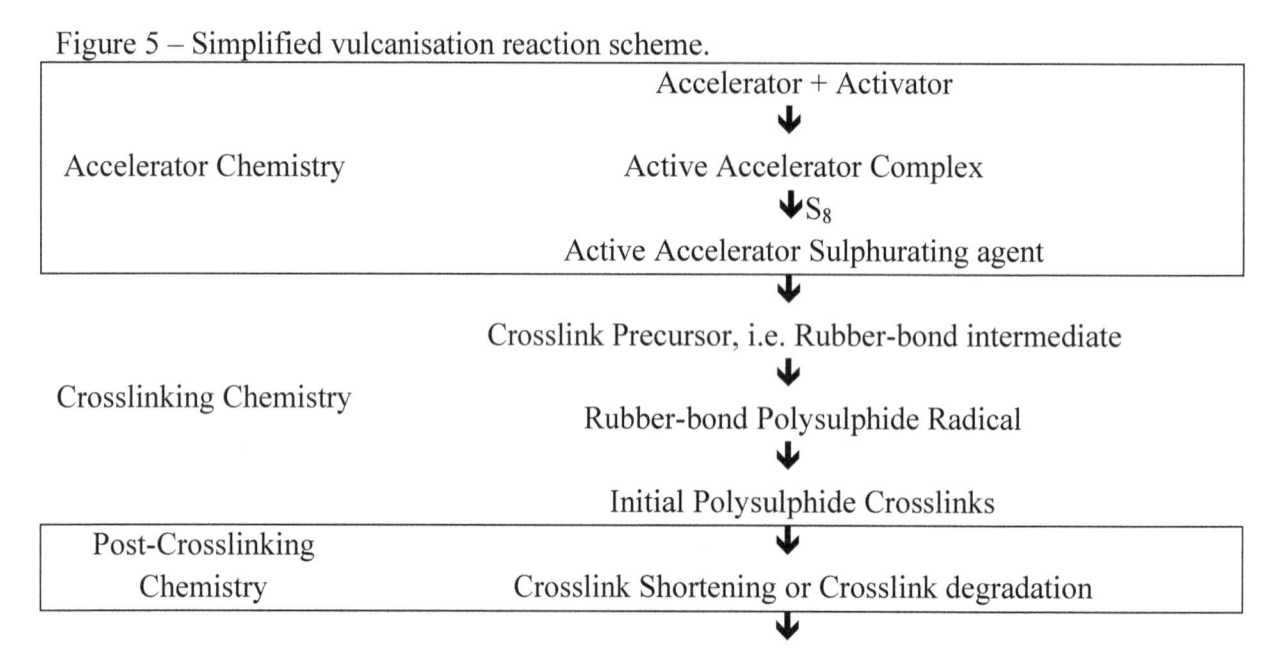

Figure 5 – Simplified vulcanisation reaction scheme.

Accelerator Chemistry	Accelerator + Activator ↓ Active Accelerator Complex ↓S_8 Active Accelerator Sulphurating agent
	↓
Crosslinking Chemistry	Crosslink Precursor, i.e. Rubber-bond intermediate ↓ Rubber-bond Polysulphide Radical ↓ Initial Polysulphide Crosslinks
Post-Crosslinking Chemistry	↓ Crosslink Shortening or Crosslink degradation

↓

Final Vulcanizate Network

Source: COSTA et al., 2003; GHOSH et al., 2003; APREM; JOSEPH; THOMAS, 2005; ANANDHAN et al., 2012.

A detailed discussion on the different reaction mechanisms for the different vulcanisation systems can be found in studies from Debnath and Basu (1996), Nieuwenhuizen (1999; 2001), Gradwell and Groof (2001), Susamma, Mini and Kuriakose (2001), Coran (2003; 2005), Ghosh et al. (2003), Marykutty et al. (2003), Aprem, Joseph and Thomas

(2005), Sahoo et al. (2007), Alam, Mandal and Debnath (2012a; 2012b), Anandhan et al. (2012), and Dondi et al. (2015).

Figure 6 – Generalised steps for vulcanisation.

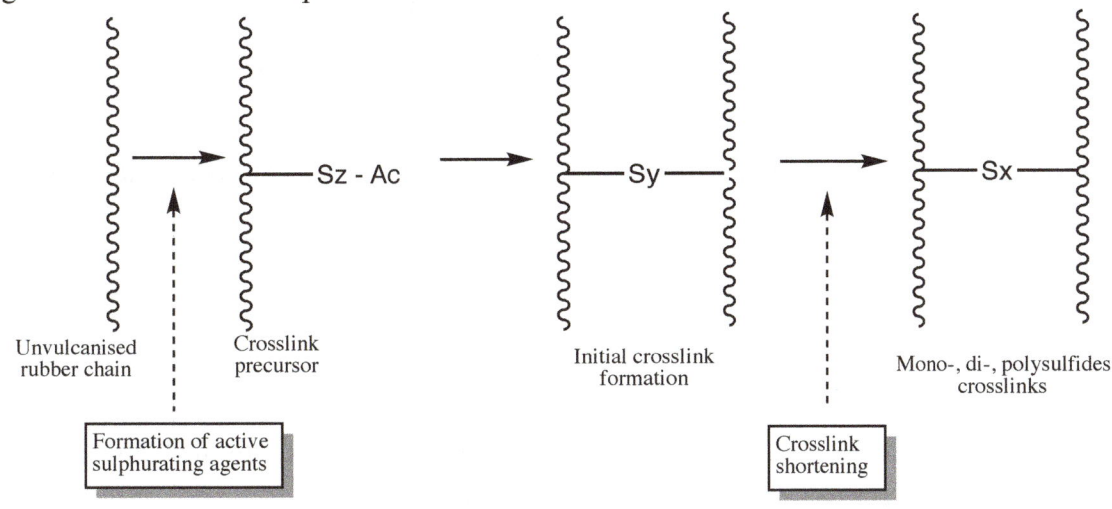

Source: NIEUWENHUIZEN, 2001 (adapted).

Sulphur crosslinks comprise various chemical structures. They may be present as mono-, di-, or polysulphides, cyclic sulphides and pendent groups (Figure 7), these structural differences have important effect on the mechanical properties of the final rubber articles (CHOI, 2000; FAN et al., 2001; CORAN, 2003; APREM; JOSEPH; THOMAS, 2005; CORAN, 2005; CHOI; KIM, 2015; HERNÁNDEZ et al., 2015).

Figure 7 – Different types of crosslinks.

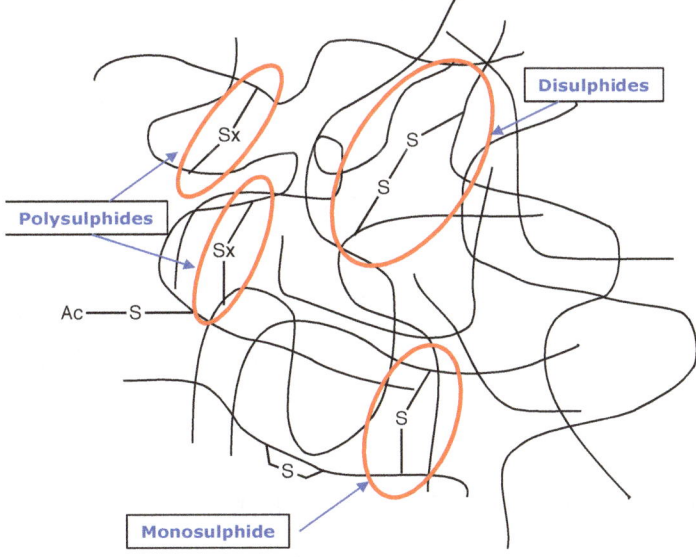

Source: CORAN, 2003 (adapted)

The choice of the accelerator and the accelerator/sulphur ratio dictates the type and the crosslink density of the final crosslink network (CORAN, 2003; APREM; JOSEPH; THOMAS, 2005; GONZÁLEZ et al., 2005a). Usually, vulcanisation systems are divided according to their types with respect to the lengths of the crosslinks formed and are categorised as conventional vulcanisation (CV), semi-efficient vulcanisation (SEV), and efficient vulcanisation (EV) (FRANTA, 1989; HERNÁNDEZ et al., 2015).

Efficient vulcanisation (EV) systems employ high amounts of accelerators and low amounts of sulphur, or utilise sulphurless formulations, yielding high levels of monosulphides crosslinks, which are more thermally stable. (FRANTA, 1989; DATTA, 2002; CORAN, 2003; GONZÁLEZ et al., 2005a; DIJKHUIS; NOORDERMEER; DIERKES, 2009; MANSILLA et al., 2015; MOVAHED; ANSARIFAR; MIRZAIE, 2015). However, greater amounts of accelerators also increase the amounts of pendent groups that are generated through this process (CORAN, 2003; GHOSH et al., 2003).

Moreover, conventional vulcanisation (CV) requires high levels of elemental sulphur and lower levels of accelerator, leading to more polysulphide crosslinks, whose dynamic properties (e.g. dynamic fatigue resistance) are superior but possess lower thermal resistance (FRANTA, 1989; DATTA, 2002; DIJKHUIS; NOORDERMEER; DIERKES, 2009; MANSILLA et al., 2015). Table 4 compares the different vulcanisation systems and the amounts of the different types of sulphide crosslinks achieved.

Table 4 – Different vulcanisation systems.

	Accelerator-Sulphur ratio	Poly- and disulphide crosslinks	Monosulphide crosslinks (%)	Heat ageing resistance
Conventional Vulcanisation (CV)	≤ 0.7	95%	5%	Low
Semi-efficient vulcanisation (SEV)	0.8 – 2.5	50%	50%	Medium
Efficient Vulcanisation (EV)	> 2.5	20%	80%	High

Source: FRANTA, 1989; DATTA, 2002; APREM; JOSEPH; THOMAS, 2005; MARZOCCA; MANSILLA, 2007.

Nevertheless, the accelerator/sulphur ratio is more sensitive to NR than to other rubbers, e.g., polybutadiene (BR) and styrene-butadiene rubber (SBR) (CORAN, 2005; MARZOCCA; MANSILLA, 2007). Generally, sulphur levels are related to the overall crosslink density achieved through the process, whereas the accelerator levels determine the lengths of the sulphur crosslinks (MARZOCCA; MANSILLA, 2007).

As previously mentioned, the curing process can occur via different means than sulphur crosslinking, e.g., via metallic oxide. Carboxylated nitrile rubber is composed of carboxylated groups, in addition to unsaturated units and nitrile groups. For this reason, this type of nitrile rubber is able to form ionic crosslinks between the carboxylic groups and metallic oxides, e.g., zinc oxide (IBARRA; MARCOS-FERNÁNDEZ; ALZORRIZ, 2002; NIYOGI, 2007; IBARRA; RODRÍGUEZ; MORA-BARRANTES, 2008). Nonetheless, these crosslinks are less thermal resistant (IBARRA; RODRÍGUEZ; MORA-BARRANTES, 2008). Figure 8 shows the crosslinking interaction between the carboxylic groups and zinc ions.

Figure 8 – Interaction of Zn(II) with carboxyl groups from carboxylated nitrile rubber (XNBR).

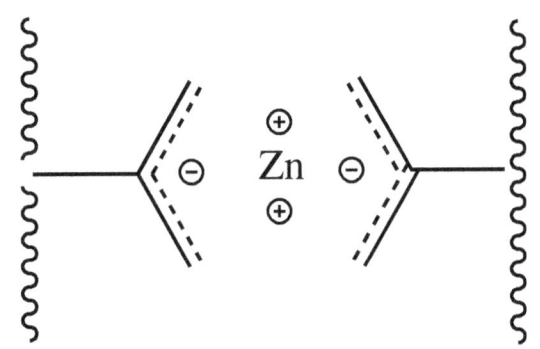

Source: IBARRA; RODRÍGUEZ; MORA-BARRANTES, 2008 (adapted).

1.4.3 Vulcanisation kinetics

Kinetic evaluations of crosslinking process are useful for understanding and modelling the effects of the formulations employed on the curing behaviour of rubbers. The assessment of rheometric parameters, i.e., the torque changes, is one of the several techniques described in the literature, which has been used to obtain the cure rate constant (k), reaction order (n), and the activation energy (E_a). According to Chough and Chang (1996), the torque change can be directly related to crosslink formation.

From torque values, the conversion (x) can be obtained as a function of time, using Equation 1 (DICK; PAWLOWSKI, 1996).

$$x_t = \frac{(C_t - C_L)}{(C_H - C_L)} \tag{1}$$

where x_t is the conversion variable at a given time, C_t is the torque at a given time; C_H is the maximum torque, and C_L is the minimum torque.

The cure rate constant (k), and the reaction order (n) can be calculated from the conversion-time curve for time values at/or superior to those with maximum cure rate. The kinetic equation is given in Equation 2 (DICK; PAWLOWSKI, 1996).

$$\frac{d_x}{d_t} = (1 - x)^n \tag{2}$$

where d_x/d_t is the conversion ratio, n is the reaction order, and t is the time.

For first-order kinetic reactions, n is equal to 1. The integrated form of the equation results in Equation 3, from which the cure rate constant (k) can be obtained through a linear regression.

$$\ln(1 - x) = k(t - t_i) \tag{3}$$

where t_i is the induction time, i.e., scorch time (t_{s1}).

However, when the reaction order is different from 1, the integrated form of Equation 2 results in Equation 4, from which k and n can be obtained by employing iterative methods.

$$(1 - n)^{-1}(1 - x)^{(1-n)} = \frac{1}{(1-n)} - k(t - t_i) \tag{4}$$

The reaction rate and temperature have an exponential relationship. Thus, an Arrhenius equation (Equation 5) can be used to estimate the activation energy (E_a). According to Prime (1981), these simple mathematical models satisfactorily fit the kinetics of most rubber vulcanisation cure systems.

$$k = Ae^{\left(-E_a/RT\right)}$$
(5)

where E_a is the activation energy, R is the gas constant, and T is the absolute temperature in Kelvin.

1.4.4 Degradation process

Every polymeric material is subjected to ageing during its use, which is caused by different factors, e.g., heat, humidity, oxidation, and aggressive media (MOSTAFA et al., 2009; WOO; PARK, 2011; XIONG et al., 2013). The use of nitrile rubber in the automotive industry is attractive; however, its resistance to thermal ageing can be limited, because of the presence of unsaturated bonds in the butadiene units of the polymer backbone (DELOR-JESTIN et al., 2000).

During thermal-oxidative ageing, modifications to the polymer chain and crosslink network can occur, leading to detrimental effects on the mechanical properties of the materials. (MOSTAFA et al., 2009; ZHAO et al., 2013). Thus, the addition of anti-degrading components to the formulation, i.e., antioxidants and antiozonants, is often necessary. These additives act to retain the useful properties of the rubber articles (BARLOW, 1988; FRANTA, 1989; DATTA et al., 2007).

Zhao et al. (2013) demonstrated that acrylonitrile content could influence the thermal stability of NBR samples. However this can only be considered true because higher acrylonitrile content correspond to lower butadiene content. Pazur, Cormier and Korhan-Taymaz (2014), Celina et al. (1998), and Delor-Jestin et al. (2000) showed that acrylonitrile units are not involved in oxidation reactions. In addition, Pazur, Cormier and Korhan-Taymaz (2014) found that higher acrylonitrile content corresponded to lower oxidation activation energies, but found that the same species were formed regardless the acrylonitrile content.

Rubber thermo-ageing is marked by the initial consumption of unsaturated bonds (DELOR-JESTIN et al., 2000; ZHAO et al., 2013; PAZUR; CORMIER, 2014), followed by the formation of alkyl and/or alkoxy radicals (ZHAO et al., 2013), which leads to the formation of hydroxyl, carbonyl, and ester products (CELINA et al., 1998; DELOR-JESTIN et al., 2000; ZHAO et al., 2013; PAZUR; CORMIER, 2014) in addition to polymer main chain scission (ZHAO et al., 2013) and zinc carboxylates formation (DELOR-JESTIN et al., 2000; CHOI; KIM, 2012) . A simplified scheme of this oxidation route is depicted in Figure 9 and Figure 10. A more detailed description of these oxidation reactions is given by Pazur, Cormier and Korhan-Taymaz (2014).

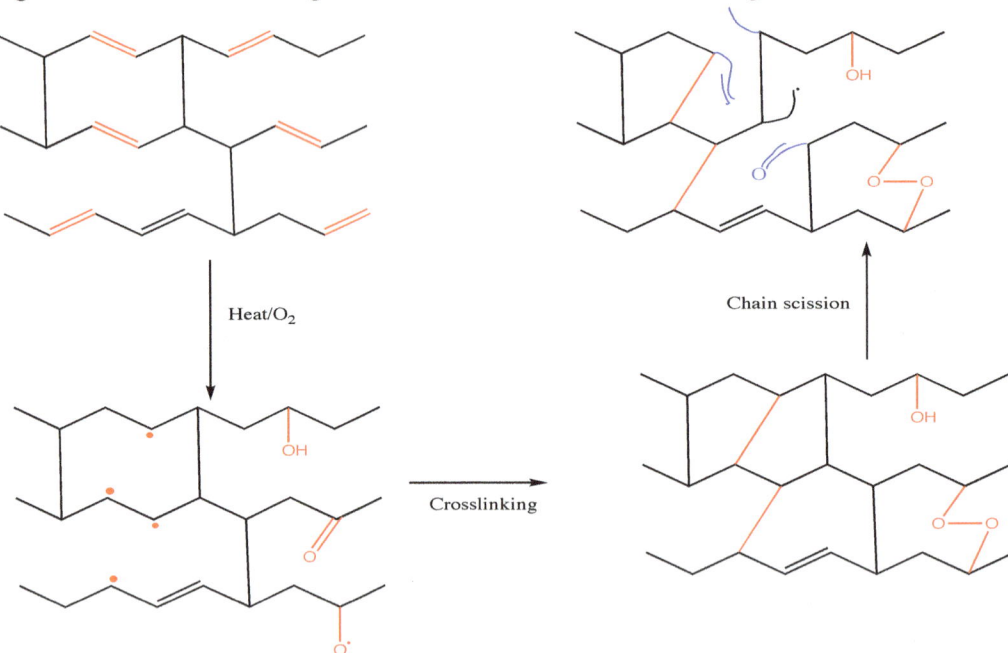

Figure 9 – Chemical changes observed in nitrile rubber during thermal oxidation.

Source: ZHAO et al., 2013 (adapted)

Crosslink networks are also severely affected by the thermal ageing process (CHOI; KIM, 2012; YANG; ZHAO; LIU, 2013; ZHAO et al., 2013). Sulphide crosslinks, mainly polysulphides, are cleaved and rearranged via this process (EL-NEMR, 2011; CHOI; KIM, 2012). Moreover, extra crosslinks are formed, which strongly compromises the mechanical properties of the products.

The loss of additives, e.g., plasticisers, is also noted during the ageing process (BUDRUGEAC, 1995; DELOR-JESTIN et al., 2000; YANG; ZHAO; LIU, 2013; ZHAO et al., 2013), which leaves the rubber materials more exposed to thermal degradation (ZHAO et al., 2013). Nevertheless, Budrugeac (1995) did not observe the loss of additives under pressure tests.

A stiffening of the rubber is usually observed, and this observation is connected to a loss of plasticisers and to an increase in the crosslink density (WOO; PARK, 2011; ZHAO et al., 2013; PAZUR; CORMIER, 2014); thus as the material becomes harder and more brittle, a premature failure of the material occurs. Furthermore, according to Celina et al. (1998), and Mostafa et al. (2009), carbon black strongly affected the oxidation stability of the rubber compounds. The filler hastened the absorption of oxygen, which catalysed the rubber oxidation.

Figure 10 – Oxidation reactions of elastomers.

$$RH \rightarrow R\bullet + H\bullet$$
$$R + O_2 \rightarrow ROO\bullet$$
$$ROO\bullet + RH \rightarrow ROOH + R\bullet$$
$$2\ ROOH \rightarrow RO\bullet + ROO\bullet + H_2O$$
$$ROOH \rightarrow RO\bullet + \bullet OH$$
$$2\ R\bullet \rightarrow R\text{–}R$$
$$R\bullet + ROO\bullet \rightarrow ROOR$$
$$2\ ROO\bullet \rightarrow \text{(stable products)}$$

Source: BARLOW, 1988; DATTA et al., 2007; CORAN, 2003.

The mechanical properties of rubber articles are clearly affected by oxidation, owing to the pronounced chemical changes observed. Zhao et al. (2013) observed small changes in low tensile moduli after NBR samples were aged at 60°C–100°C, however, at higher temperatures (125°C), the changes were more pronounced, as indicated by the increase in the crosslink density of the rubber samples. Mostafa et al. (2009) observed a decrease in the moduli at 300%, and in the elongation at break after 48 h of thermal ageing at different temperatures. According to Budrugeac (1995), thermo-mechanical degradation of nitrile rubber followed a first order kinetic equation.

Regarding the chemical resistance of elastomers to different fluids, a study must be specific to the fluid of interest, and the resistance of nitrile rubber to biodiesel was already addressed in section 1.3.

Hiltz, Morchat and Keough (1993) employed dynamic mechanical thermal analyses (DMTA) to assess the fuel resistance of different elastomers and observed decreases in the low temperature properties of the rubber samples after immersing them in different fuels. Although carbon black could induce thermo-oxidation of rubber goods, Magryta, Dębek, and Dębek (2006) showed that, in high concentrations, the filler-filler interaction could prevent the rubber from swelling when immersed in oils.

2 AIMS

2.1 General aims

This Thesis aimed to develop a new nitrile rubber sulphur-based formulation with different accelerators and to assess its resistance to pure soybean biodiesel.

2.2 Specific aims

- Assess the influence of the acrylonitrile content on the nitrile rubber resistance to pure biodiesel;
- Assess the influence of different types of crosslink networks on nitrile rubber resistance to pure biodiesel;
- Assess the influence of different sulphur-based curing systems on nitrile rubber resistance to biodiesel;
- Assess the influence of different accelerators on nitrile rubber resistance to biodiesel;
- Assess the influence of a novel binary curing system formulation on the properties of nitrile rubber.

3 MATERIALS AND EQUIPMENT

3.1 Part I – Preliminary studies: The influence of acrylonitrile content and different types of crosslink networks

3.1.1 Materials

- Nitrile rubber with 33% of acrylonitrile content (N615B) – Nitriflex S/A Indústria e Comércio;
- Nitrile rubber with 45% of acrylonitrile content (N206) – Nitriflex S/A Indústria e Comércio;
- Carboxylated nitrile rubber with 28% of acrylonitrile content, and 7% of carboxylic content (Nitriclean 3350X) – Nitriflex S/A Indústria e Comércio;
- Biodiesel from soybean oil – Centro de Pesquisa e Desenvolvimento Leopoldo Américo Miguez de Mello (CENPES/PETROBRAS);
- Carbon black (Spheron 6630) – Cabot Brasil Indústria e Comércio Ltda.;
- N-tert-butyl 2-benzothiazole sulfenamide (TBBS) – commercial grade;
- Stearic Acid – commercial grade;
- Sulphur – commercial grade;
- Zinc Oxide – commercial grade;

3.1.2 Equipment

- Dynamometer, from EMIC, model DL2000. Located at: Universidade do Estado do Rio de Janeiro (UERJ), Brazil.
- Hydraulic press, from Parabor Ltda. Located at: Universidade do Estado do Rio de Janeiro (UERJ), Brazil.
- Moving Die Rheometer (MDR), MDPt, from TechPro. Located at: Universidade do Estado do Rio de Janeiro (UERJ), Brazil.
- Open mill roll, from Parabor Ltda. Located at: Universidade do Estado do Rio de Janeiro (UERJ), Brazil.
- Durometer Shore A, from Parabor Ltda. Located at: Universidade do Estado do Rio de Janeiro (UERJ), Brazil
- Scanning Electron Microscope, JSM-6510LV, from JEOL. Located at: Universidade do Estado do Rio de Janeiro (UERJ), Brazil.

3.2 Part II – Formulation development: The influence of binary sulphur-based curing systems

3.2.1 <u>Materials</u>

- Nitrile rubber with 45% of acrylonitrile content (N206) – Nitriflex S/A Indústria e Comércio;
- Biodiesel from soybean oil – ASG Analytik-Service Gesellschaft mbH (purchased)
- Carbon black (Spheron 6630) – Cabot Brasil Indústria e Comércio Ltda;
- Stearic Acid – commercial grade;
- Sulphur – commercial grade;
- Tetramethylthiuram disulphide (TMTD) – commercial grade;
- N-cyclohexyl-2-benzothiazole sulphenamide (CBS) – commercial grade;
- Zinc Oxide – commercial grade;

3.2.2 <u>Equipment</u>

- 3D Laser Scanning Microscope, VK-X100, from Keyence. Located at: Bundesanstalt für Materialforschung und –prüfung (BAM), Germany.
- Differential scanning calorimeter, DSC/SDTA 821e, from: Mettler Toledo DSC 1 Star System. Located at: Universität Bayreuth, Germany.
- Dynamometer, from EMIC, model DL2000. Located at: Universidade do Estado do Rio de Janeiro (UERJ), Brazil.
- Fourrier transform infrared spectrometer, Varian 640-IR, from Varian. Located at: Universidade do Estado do Rio de Janeiro (UERJ), Brazil.
- Hydraulic press, from Parabor Ltda. Located at: Universidade do Estado do Rio de Janeiro (UERJ), Brazil.
- Moving Die Rheometer (MDR), MDPt, from TechPro. Located at: Universidade do Estado do Rio de Janeiro (UERJ), Brazil.
- Nuclear magnetic resonance spectrometer, VNMRS-500Hz, from Varian. Located at: Universidade do Estado do Rio de Janeiro (UERJ), Brazil.
- Open mill roll, from Parabor Ltda. Located at: Universidade do Estado do Rio de Janeiro (UERJ), Brazil.

- Rheometic Scientific Analyser, RDA III, from TA Instruments. Located at: Universität Bayreuth, Germany.
- Scanning Electron Microscope, JSM-6510, from JEOL. Located at: Universität Bayreuth, Germany.
- Universal Hardness Sensor, from Zwick GmbH & Co. KG. Located at: Universität Bayreuth, Germany.
- Universal Testing Machine Z050, from Zwick GmbH & Co. KG. Located at: Universität Bayreuth, Germany.

4 METHODS

4.1 Part I – Preliminary studies: The influence of acrylonitrile content and different types of crosslink networks

4.1.1 Compounding

Different compositions were prepared using different types of nitrile rubber samples. Three different samples were used: a nitrile rubber sample with 33% of acrylonitrile content, a nitrile rubber sample with 45% of acrylonitrile content, and a carboxylated nitrile rubber sample with 28% of acrylonitrile content. For identification purposes the compositions were labelled as shown in Table 5.

Table 5 – Identification for each composition obtained from different nitrile rubber samples.

Composition label	Rubber sample used
NBR33	Nitrile rubber sample with 33% of acrylonitrile content
NBR45	Nitrile rubber sample with 45% of acrylonitrile content
XNBR	Carboxylated nitrile rubber sample with 28% of acrylonitrile content and 7% of carboxyl group content

The compositions were prepared in a roll mill at 50°C±5°C. The compositions were processed according to the standard formulation given by the American Society for Testing and Material (ASTM) D 3187:11 (Table 6).

Prior to the addition of the others ingredients, the elastomer was subjected to a mastication process, which consisted of banding the raw elastomer on the mill such that the elastomer became easier to be processed.

The addition order of the materials was determined by the ASTM standard. Stearic acid and zinc oxide, which were were previously mixed, were added first. These materials were followed by the addition of sulphur and TBBS (N-tert-butyl 2-benzothiazole sulphenamide), which had also been previously mixed. Carbon black (filler) was the last to be added.

After incorporating each ingredient, horizontal cuts that were ¾ of the width, were made three times on each side of the batch. If any amount of the ingredients was noticed to have been dropped through the mill, it was returned to the batch until full incorporation of the ingredients was achieved.

Table 6 – Standard formulation recipe for nitrile rubber compositions according to ASTM D3187:11

Component	Function	phr[a]
Nitrile Rubber	Elastomer	100
Zinc Oxide	Activator	3
Stearic Acid	Activator	1
Sulphur	Cure Agent	1.5
TBBS[b]	Accelerator	0.7
Carbon Black	Filler	40

Footnote: (a) – parts per hundred of rubber; (b) – N-tert-butyl-2-benzothiazole sulphenamide.

4.1.2 Vulcanisation

The vulcanisation time (optimum cure time) for each composition was obtained using a moving die rheometer.

Rheometric tests were conducted according to ASTM D 5289:12. A sample from each composition was analysed for 1 hour at 160°C. The analyses were conducted for no longer than 24 hours after the preparation of the composition.

Vulcanisation was conducted in a hydraulic press at 160°C according to ASTM D3182.

4.1.3 Vulcanisation kinetic

Rheometric parameters for each composition were obtained in a Moving Die Rheometer, MDPt (TechPro) under isothermal conditions. The compositions were tested at three different temperatures: 140°C, 150°C and 160°C. The maximum torque (C_H), minimum torque (C_L), scorch time (t_{s1}) and optimum vulcanisation time (t_{90}) were assessed for each sample.

These parameters were used to calculate the cure rate constant (k), reaction order (η), and cure activation energy (E_a) of each composition, employing the Equations 1, 3, 4, and 5, which were described in section 1.4.3.

4.1.4 Crosslink density

Three small square samples were cut from each composition and weighed in air and in acetone to calculate the initial mass and the density of the composition.

Afterwards, the square samples were immersed in acetone for at least one week at room temperature to assure they reached their swelling equilibrium.

The swollen samples were weighed and then kept under vacuum for at least 12 hours before being weighed again.

The crosslink densities of the compositions were calculated using the Flory-Rehner equation, as depicted in Equation 6 (EL-NEMR, 2011). The crosslink densities were calculated by discounting the amount of carbon black in the compositions.

$$\mu = \frac{-[\ln(1-v_r) + v_r + \alpha \cdot v_r^2]}{\left[V_0\left(v_r^{\frac{1}{3}} - \frac{v_r}{2}\right)\right]} \qquad (6)$$

where μ is the crosslink density, v_r is the volume fraction of rubber in equilibrium of the swollen vulcanised sample, V_0 is the molar volume of the solvent (73.40 mL.mol^{-1}), and α is the interaction parameter between the solvent and the elastomer (0.345 for acetone-nitrile rubber).

4.1.5 Immersion tests

The immersion tests followed the procedures described in ASTM D 471:12 and were conducted in an oven with forced air circulation at 100°C for 22 hours. The oven was kept closed during the entire test. All test specimens were kept completely immersed throughout the test.

Pure ethylic biodiesel obtained from soybean oil was used, and its properties were in accordance with Brazilian regulations (Table 1 in section 1.1.3).

4.1.6 Change in mass

Small square specimens were cut from the vulcanised sheets to assess changes in their masses after immersion. The specimens were weighed in air on a balance with an accuracy of 0.1 mg.

After cooling the specimens as described in ASTM D 471:12, the immersed samples had their surfaces dried with filter paper and were immediately weighed. The results were the average change in mass from the test specimens. Changes in mass were calculated according to Equation 7.

$$\Delta M, \% = \frac{M_2 - M_1}{M_1} \times 100 \qquad (7)$$

where ΔM is the change in mass after immersion (%), M_1 is the initial mass, and M_2 is the final mass.

4.1.7 Mechanical tests

4.1.7.1 Strain-stress

Stress-strain tests were performed on a dynamometer according to ASTM D 412:13. The tested specimens were placed in dynamometer grips with crosshead speed of 500 mm/min. The final result was the average of five measurements.

4.1.7.2 Tear strength

Tear strength tests were conducted on a dynamometer according to ASTM D 624:12. The tested specimens were placed in dynamometer grips with crosshead speed of 500 mm/min. The final result was the average of five measurements.

4.1.7.3 Hardness

Hardness tests were conducted according to ASTM D 2240:10 using a Shore A durometer. Five measurements were conducted on the test specimens, which were approximately 6 mm thick. The final result was the average of the five measurements.

4.1.8 Scanning Electron Microscopy (SEM)

The morphological aspects of the fracture surface of the samples were analysed by scanning electron microscopy (SEM). After the mechanical tests, small pieces of the immersed and non-immersed tensile strength test specimens were sputtered with a gold film to allow electric conduction. SEM analyses were conducted using an electron beam with a 15 kV acceleration voltage.

4.2 Part II – Formulation development: The influence of binary sulphur-based curing systems

4.2.1 Compounding

Different compositions were prepared using different formulations, employing a two-level design of experiments with a central point (2^3+3). Different amounts of sulphur and two different accelerators were used. The other components of the formulation were kept constant for all the compositions. The two accelerators used for the experiments were tetramethylthiuram disulphide (TMTD), and N-cyclohexyl-2-benzothiazole sulphenamide (CBS). The amounts of each component in respect to each level of the design of experiments are presented in Table 7.

Table 7 – Amount in phr[a] with respect to their coded amount.

Coded amount	TMTD[b]	CBS[c]	Sulphur
-1	1	0	0.5
0	2	1	1
+1	3	2	1.5

Footnote: (a) – Parts per a hundred parts of rubber; (b) – Tetramethylthiuram disulphide; (c) – N-cyclohexyl-2-benzothiazole sulphenamide.

The formulation of each composition is presented in Table 8. The compositions were labelled according to the amounts in phr (part per a hundred parts of rubber) of each vulcanisation component employed in each formulation following the order: TMTD/CBS/sulphur. All the formulations follow this notation throughout this Thesis.

The compositions were prepared in a roll mill at 50°C±5°C. Prior to the addition of the others ingredients, the elastomer was subjected to a mastication process, which consisted of banding the raw elastomer on the mill such that the elastomer could be more readily processed.

The addition order of the components was as follows: stearic acid and zinc oxide were mixed then added first. They were followed by the addition of half of the amount of carbon black. After the incorporation of this portion, horizontal cuts that were ¾ of the width were made three times on each side of the batch. The remaining amount of carbon black was added afterwards and horizontal cuts, ¾ of the width, were made again three times on each side of the batch. The accelerators (TMTD and CBS), previously mixed, were also added together after this process and sulphur was added last.

Table 8 – Design of experiments (2^3+3) for nitrile rubber with 45% of acrylonitrile content. Between brackets indicate the amount in phr of each curing system component of the formulations (TMTD/CBS/sulphur).

	1 (1/0/0.5)	2 (1/0/1.5)	3 (1/2/0.5)	4 (1/2/1.5)	5 (3/0/0.5)	6 (3/0/1.5)	7 (3/2/0.5)	8 (3/2/1.5)	9* (2/1/1)
				Coded amount					
TMTD[a]	-1	-1	-1	-1	+1	+1	+1	+1	0
CBS[b]	-1	-1	+1	+1	-1	-1	+1	+1	0
Sulphur	-1	+1	-1	+1	-1	+1	-1	+1	0
				Amount in phr[c]					
Nitrile rubber	100	100	100	100	100	100	100	100	100
Zinc Oxide	3	3	3	3	3	3	3	3	3
Stearic Acid	1	1	1	1	1	1	1	1	1
Carbon black	40	40	40	40	40	40	40	40	40
TMTD[a]	1	1	1	1	3	3	3	3	2
CBS[b]	0	0	2	2	0	0	2	2	1
Sulphur	0.5	1.5	0.5	1.5	0.5	1.5	0.5	1.5	1

Footnote: (a) – Tetramethylthiuram disulphide; (b) – N-cyclohexyl-2-benzothiazole sulphenamide; (c) – parts per a hundred parts of rubber. *repeated three times.

4.2.2 Vulcanisation

The vulcanisation time (optimum cure time) for each composition was obtained in a moving die rheometer.

The rheometric tests were conducted according to ASTM D 5289:12. The sample from each composition was analysed for 1 hour at 160°C. The analyses were conducted no longer than 24 hours after the compositions were prepared.

Vulcanisation was conducted in a hydraulic press at 160°C, according to ASTM D 3182:15.

4.2.3 Crosslink density

The crosslink density of each composition was calculated using the same methodology described in section 4.1.4 using the Flory-Rehner equation (EL-NEMR, 2011) (Equation 6) at room temperature and using acetone as a solvent. The crosslink densities were calculated by discounting the amount of carbon black in the compositions.

4.2.4 Ageing tests

4.2.4.1 Ageing in air

Heat ageing tests were conducted to assess the effect of temperature on the rubber degradation process. The specimens were kept in an oven with forced air circulation at 100°C for 22 h according to ASTM D 573:10. Strain-stress tests (ISO 37:05) were conducted after the specimens were cooled to room temperature for no less than 16 h and no more than 96 h.

4.2.4.2 Ageing in biodiesel

Biodiesel ageing tests were conducted in an oven with forced air circulation at 100°C, following ASTM D 471:12. All the test specimens were kept completely immersed throughout the test.

Pure methylic soybean biodiesel was used, and its main properties are given in Table 9, which falls within the European specifications as per EN 14214:10 (Table 1). The complete list of properties of the soybean biodiesel used in these experiments is shown in Table 16 (ANNEXE A).

Table 9 – Main chemical and physical properties of soybean biodiesel (soybean methyl ester).

Property	Values	Unit
Density (15°C)	884.9	kg/m^3
Viscosity (40°C)	4.118	mm^2/s
Sulphur content	0.75	mg/kg
Acid number	0.176	mg KOH/g

Source: Informed by the supplier (ASG Analytik-Service Gesellschaft mbH).

4.2.5 Gravimetric tests

The oil uptake profiles of the rubber compositions were assessed for 700 h. Rectangular specimens (50 mm x 10 mm) with average thicknesses of 2,5 mm were cut from each composition. At least three specimens from each composition were used, and they were kept completely immersed in biodiesel throughout the test.

At given times, the specimens were removed from the oil, had their surface carefully dried, weighed and were placed back into the oil. Each specimen was kept out of the oil for less than 10 minutes. The specimens were weighed in air on a balance with an accuracy 0.1 mg. The changes in mass were calculated according to Equation 2 (section 4.1.6).

4.2.6 Stress-strain

Stress-strain tests were performed on a universal testing machine according to ISO 37. The tested specimens were placed in the dynamometer grips with crosshead speed of 200 mm/min. The final result was the average of five specimens.

Stress-strain tests were also conducted with heat-aged and biodiesel-aged samples after 22 h of ageing.

4.2.7 Hardness

Hardness tests were conducted according to ASTM D 2240:10 using a Universal Hardness Sensor. Five measurements were made on the test specimens, which were approximately 6 mm thick. The final result was the average of these five measurements.

Hardness tests were also conducted with biodiesel-aged samples after 22 h of immersion.

4.2.8 Differential scanning calorimetry (DSC)

The glass transition temperatures (T_g) of the samples were evaluated by differential scanning calorimetry (DSC). The temperature range of the tests was from -60°C to 40°C, with a heating, and cooling rate of 10°C/min, under a flow of nitrogen.

After the first heating run, the samples were held isothermally at 40°C for 2 minutes; then, they were cooled to -60°C and held isothermally for 5 minutes. Afterwards, the second heating run was conducted at a temperature of up to 40°C.

The T_g was calculated as the temperature of the maximum point from the first derivative curve of the second heating run.

The tests were performed with non-aged specimens and with biodiesel-aged specimens, which were immersed for 22 h, 46 h, 166 h, and 700 h.

4.2.9 Dynamic mechanical thermal analysis (DMTA)

Dynamic mechanical thermal analyses (DMTA) were performed in torsion mode in a temperature range from -50°C to +50°C with a heating range of 3°C/min and 1Hz frequency.

Analyses were conducted with non-aged specimens and with biodiesel-aged specimens, which were immersed for 166 h and 700 h. The immersed samples were removed from the oil and their surfaces were dried; and they were then immediately analysed without extracting the absorbed oil.

4.2.10 Scanning electron microscopy (SEM)

The tensioned fracture surface of the compositions were analysed by scanning electron microscopy (SEM). Both non-immersed and immersed (after 22 h) specimens were evaluated.

The non-tensioned fracture surfaces of the compositions were analysed by SEM. The specimens were cooled in liquid nitrogen and manually fractured. One part of the specimen was immediately analysed by SEM, and the other part was stored for 80 days, safe from dust, and then analysed by SEM.

The non-fractured surfaces of the compositions were also analysed by SEM to observe possible exudation of the components.

Prior to SEM analyses, the samples were sputtered with a gold film to allow electric conduction. SEM analyses were conducted using an electron beam with a 10 kV acceleration voltage.

4.2.11 Confocal Laser Scanning Microscopy (CLSM)

The morphological aspects of the fracture surfaces of the samples were analysed by confocal laser scanning microscopy (CLSM). After the mechanical tests, small pieces of the immersed and non-immersed specimens from the strain-stress tests were analysed with a 3D Laser Scanning Microscope.

4.2.12 Attenuated total reflectance Fourier transform infrared (ATR-FTIR) spectroscopy

Thin films of the compositions were analysed by attenuated total reflectance Fourier transform infrared (ATR-FTIR) spectroscopy. The spectra were collected from 600 to 4000 cm^{-1}, with a resolution of 4 cm^{-1}.

4.2.13 Nuclear magnetic resonance (NMR)

All H^1NMR spectra were acquired using $CDCl_3$ as solvent and residual protons as internal reference (7.27 ppm). Spectra were recorded at room temperature with 164K data points, 64 scans at a spectral width of 8000 Hz, relaxation delay of 1 s, and acquisition time of 2.045 s. Exponential line broadening (1.5 Hz), automatic phase correction, and baseline correction (degree of polynomial equal to 5) were applied to each spectrum.

Small specimens of the vulcanised compositions were quickly dipped in the solvent, for no longer than 10s, only to solubilise the components from the composition surfaces.

4.2.14 Statistical analyses

Statistical analyses were conducted using *Statistica 8* software and descriptive statistics with a 95% of confidence interval.

4.3 Experimental scheme

4.3.1 Part I – Preliminary studies: The influence of acrylonitrile content and different types of crosslink networks

Figure 11 – Summary of the Part I experimental section

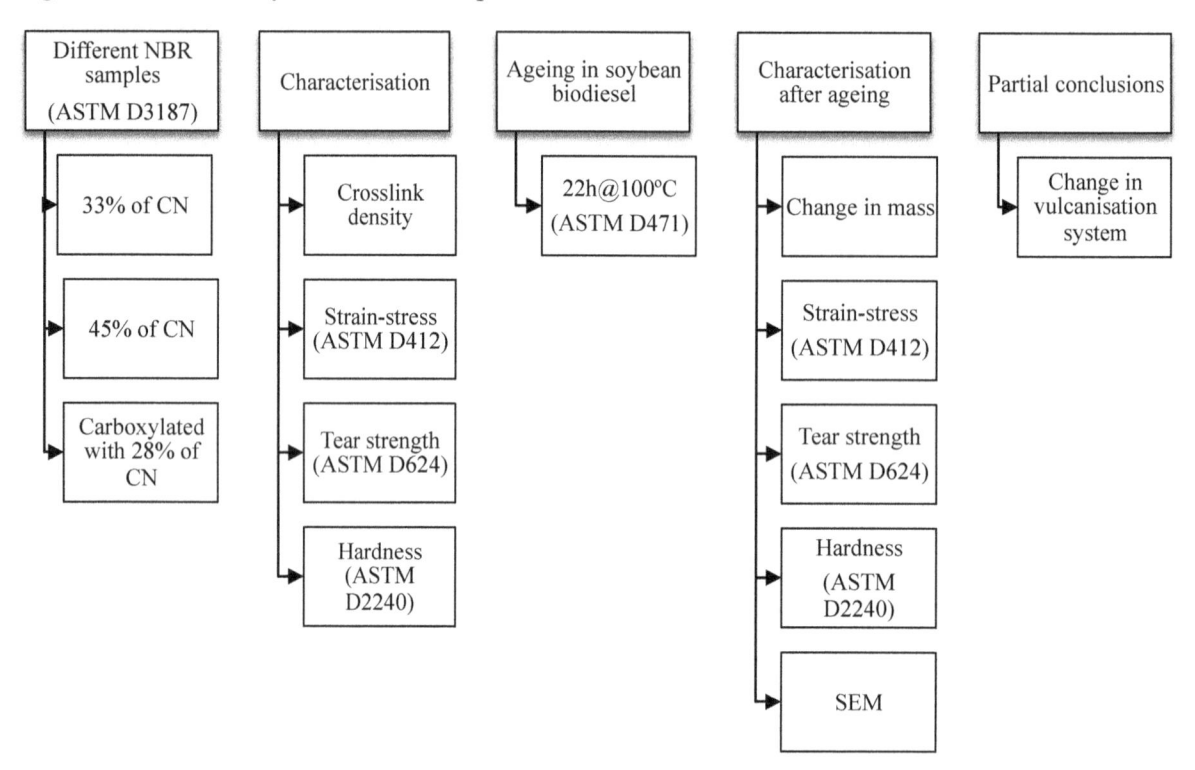

Different NBR samples (ASTM D3187)	Characterisation	Ageing in soybean biodiesel	Characterisation after ageing	Partial conclusions
33% of CN	Crosslink density	22h@100°C (ASTM D471)	Change in mass	Change in vulcanisation system
45% of CN	Strain-stress (ASTM D412)		Strain-stress (ASTM D412)	
Carboxylated with 28% of CN	Tear strength (ASTM D624)		Tear strength (ASTM D624)	
	Hardness (ASTM D2240)		Hardness (ASTM D2240)	
			SEM	

4.3.2 Part II – Formulation development: The influence of binary sulphur-based curing systems

Figure 12 – Summary of Part II experimental section

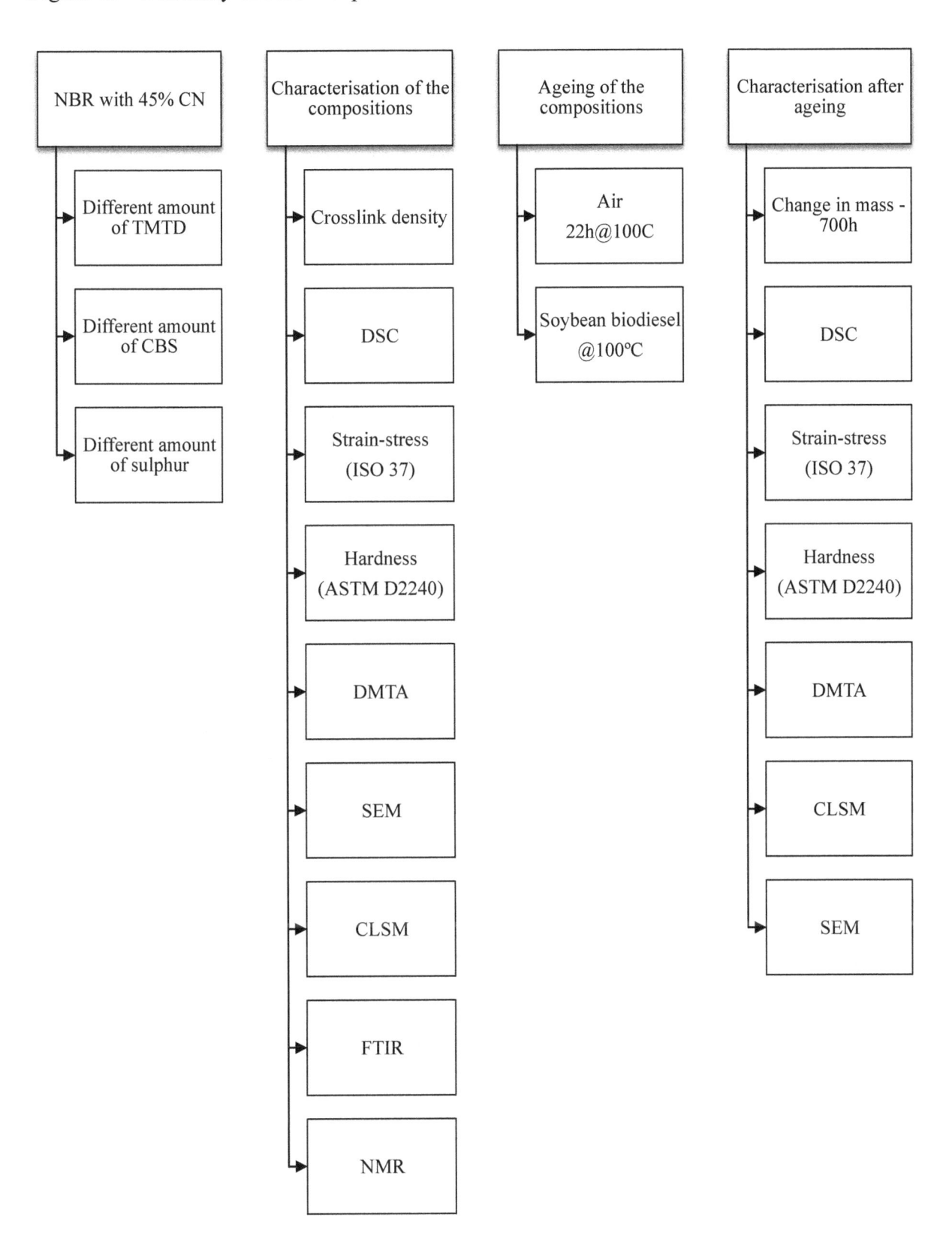

NBR with 45% CN
- Different amount of TMTD
- Different amount of CBS
- Different amount of sulphur

Characterisation of the compositions
- Crosslink density
- DSC
- Strain-stress (ISO 37)
- Hardness (ASTM D2240)
- DMTA
- SEM
- CLSM
- FTIR
- NMR

Ageing of the compositions
- Air 22h@100C
- Soybean biodiesel @100°C

Characterisation after ageing
- Change in mass - 700h
- DSC
- Strain-stress (ISO 37)
- Hardness (ASTM D2240)
- DMTA
- CLSM
- SEM

5 RESULTS AND DISCUSSION

5.1 Part I – Preliminary studies: The influence of acrylonitrile content and different types of crosslink networks

5.1.1 Characterisation of the compositions

5.1.1.1 Vulcanisation kinetics

The rheometric parameters (C_L, C_H, t_{s1}, and t_{90}) of the three compositions, which were prepared according to ASTM 3187, are presented in Table 10. Considering that the vulcanisation system employed for these compositions were the same, the differences noted should be attributed to the different elastomers structures, i.e., different acrylonitrile content or the presence of carboxyl group.

Table 10 – Rheometric parameters of the nitrile rubber compositions prepared with ASTM formulations (C_L – minimum torque; C_H – maximum torque; ΔC – maximum and minimum torques diferrence; t_{s1} – scorch time; t_{90} – optimum cure time).

Parameters	NBR33	NBR45	XNBR
Vulcanisation at 140°C			
C_L (dN.m)	2.3	1.5	1.8
C_H (dN.m)	17.2	15.8	20.6
ΔC (dN.m)	14.9	14.3	18.8
t_{s1} (min)	11.8	7.7	4.9
t_{90} (min)	31.5	41.0	42.5
Vulcanisation at 150°C			
C_L (dN.m)	2.1	1.2	2.1
C_H (dN.m)	17.1	15.6	22.3
ΔC (dN.m)	15.0	14.4	20.2
t_{s1} (min)	11.8	4.0	2.9
t_{90} (min)	19.6	36.5	27.1
Vulcanisation at 160°C			
C_L (dN.m)	1.9	1.1	1.5
C_H (dN.m)	16.7	18.4	23.7
ΔC (dN.m)	14.8	17.3	22.2
t_{s1} (min)	3.3	2.2	1.5
t_{90} (min)	11.3	32.0	23.8

The carboxylated nitrile rubber composition (*XNBR*) presented, regardless of the vulcanisation temperature, the highest maximum torque (C_H) and the highest difference between maximum and minimum torques values. It is noteworthy that the *NBR33* and *NBR45* compositions had similar behaviours. These parameters are often directly related to the density of crosslinks formed during the vulcanisation process (SOUSA et al., 2002; ELHAMOULY; MASOUD; KANDIL, 2010; AKHLAGHI et al., 2012). The higher values observed for the *XNBR* composition can be related to the higher crosslink density that was achieved during the vulcanisation process because of the presence of additional vulcanisation sites (carboxyl groups) (IBARRA; MARCOS-FERNÁNDEZ; ALZORRIZ, 2002; NIYOGI, 2007; IBARRA; RODRÍGUEZ; MORA-BARRANTES, 2008).

Moreover, the optimum cure times (t_{90}) for all the compositions obviously decreased with increases in the vulcanisation temperature.

The calculated kinetic parameters are shown in Table 11. The reaction orders (n) for all the compositions were higher than 1.0. Additionally, similar values were found for non-carboxylated compositions (*NBR33* and *NBR45*), which suggested that the vulcanisation process proceeded and involved similar chemical species. Moreover, *XNBR* vulcanisation seemed to occur via a different mechanism from *NBR33* and *NBR45*, given the different reaction order found for *XNBR*. This result implies that the vulcanisation mechanism engages different chemical species; at lower temperatures, the vulcanisation process most likely happened preferably via the carboxyl groups.

Table 11 – Cure rate constant (k), reaction order (n), and activation energy (E_a) of the nitrile rubber compositions vulcanised at different temperatures.

Compositions	Temperature (°C)	n	k ($10^3.s^{-1}$)	E_a (kJ.mol^{-1})
	140°C	1.4	3.54 ± 0.05	
NBR33	150°C	1.8	8.12 ± 0.09	100 ± 6
	160°C	1.8	13.49 ± 0.15	
	140°C	1.3	1.57 ± 0.01	
NBR45	150°C	1.6	2.42 ± 0.02	44 ± 4
	160°C	1.7	2.82 ± 0.03	
	140°C	1.1	1.06 ± 0.01	
XNBR	150°C	1.2	2.14 ± 0.02	78 ± 5
	160°C	1.5	3.02 ± 0.03	

The cure rate constants (k) for *NBR33* and for *NBR45* agreed well with their respective acrylonitrile contents, i.e., a lower acrylonitrile content correlated to a higher butadiene content and, therefore, to a greater number of reactive sites (allylic hydrogen atoms), which led to higher values of k, and lower times for t_{90}. The opposite tendency was observed for activation energies (E_a), i.e., an increase in the acrylonitrile content resulted in a decrease in E_a, implying that the nitrile groups increased the reactivity of the vicinal carbons to vulcanisation reactions.

5.1.1.2 Crosslink density

The crosslink densities of the compositions (Figure 13) were in accordance with the rheometric results shown in Table 10. The carboxylated nitrile rubber composition (*XNBR*) achieved the highest crosslink density among the compositions which were assessed, whereas *NBR33* and *NBR45* had similar results.

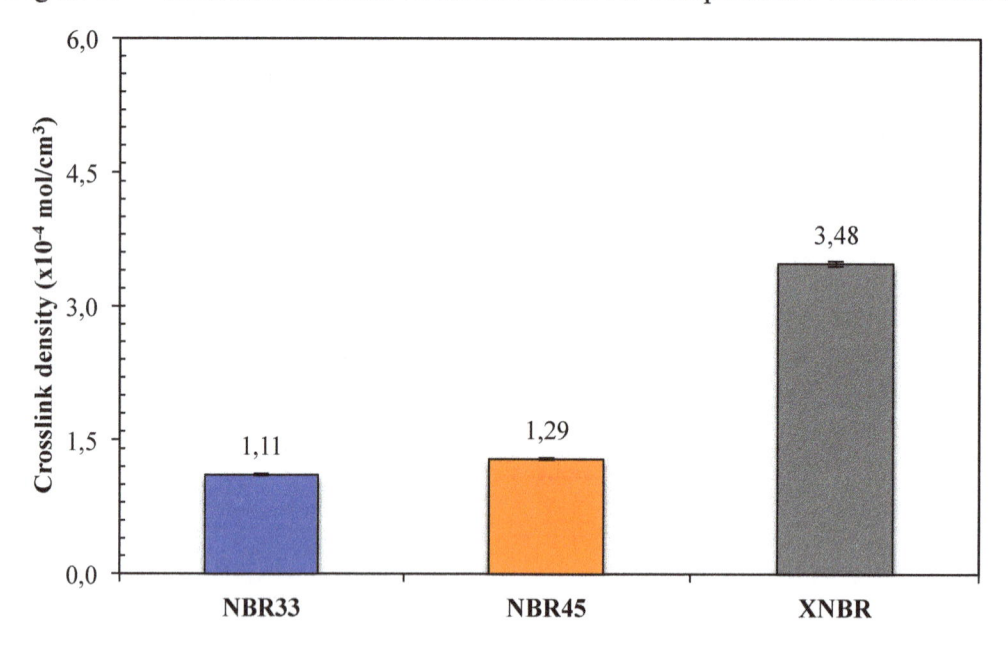

Figure 13 – Crosslink densities of the nitrile rubber compositions vulcanised at 160°C.

Composition	Crosslink density ($\times 10^{-4}$ mol/cm^3)
NBR33	1,11
NBR45	1,29
XNBR	3,48

The results obtained in this Thesis did not agree with the observations reported by Affonso and Nunes (1995), and Haseeb et al. (2010; 2011b) who suggested that an increase in acrylonitrile content would increase the crosslink density of the compositions. No relation was observed between crosslink densities with respect to acrylonitrile content. Apparently, acrylonitrile content only played a role in the kinetics of vulcanisation processes.

The obvious higher crosslink density observed for *XNBR* corroborates the higher torque difference observed during the rheometric tests, which was due to the different sites at which carboxylated nitrile rubber could be vulcanised (IBARRA; MARCOS-FERNÁNDEZ; ALZORRIZ, 2002; NIYOGI, 2007; IBARRA; RODRÍGUEZ; MORA-BARRANTES, 2008).

5.1.2 Ageing tests

5.1.2.1 Gravimetric tests

The observed changes in masses of the samples after their immersion in biodiesel are shown in Figure 14. The composition *NBR33* absorbed the largest amount of fuel; its weight increased by more than 50%; as expected, the composition with the highest acrylonitrile content (*NBR45*) swelled the least. Walker (2009) also observed a decrease in swelling as the acrylonitrile content of the rubber samples increased.

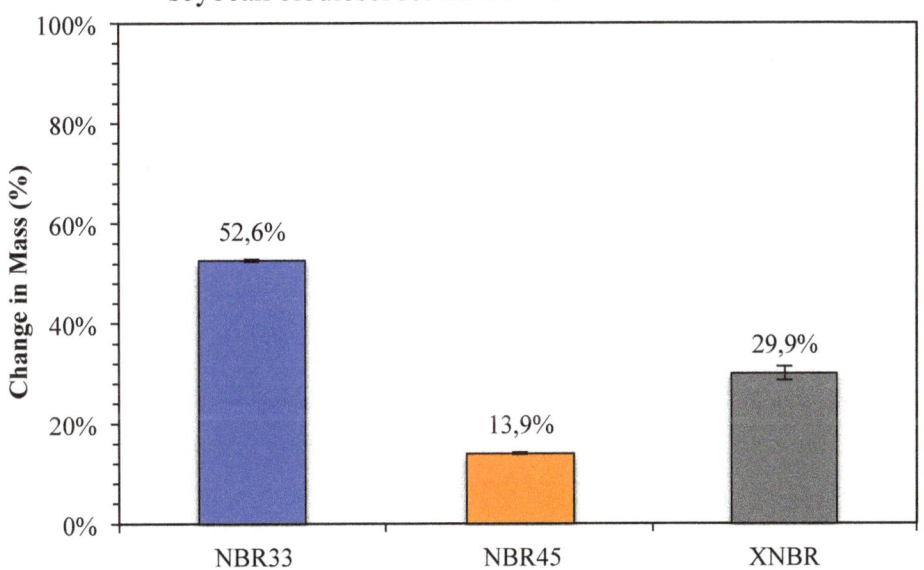

Figure 14 – Changes in mass of the nitrile rubber compositions after immersion in soybean biodiesel for 22 h at 100°C.

Composition	Change in Mass (%)
NBR33	52,6%
NBR45	13,9%
XNBR	29,9%

Nonetheless, *XNBR*, which had only 28% of acrylonitrile content, absorbed less oil than *NBR33*. The presence of carboxyl groups may have prevented a higher uptake of biodiesel by this composition.

Authors observed changes in masses as low as 3% (DUBOVSKÝ; BOŽEK; OLŠOVSKÝ, 2015), when NBR was immersed in rapeseed biodiesel B10 at 23°C, for 168 h and as high as 35% (HASEEB et al., 2010), when samples were tested in pure palm biodiesel for 500 h at 50°C.

These different results found may be associated with the different conditions of the tests. Comparing the results to observations reported in the literature must be carefully conducted because there are many parameters that must be considered (e.g., temperature, time, biodiesel origin, biodiesel concentration, and type of the nitrile rubber) and may yield significantly different results.

The biodiesel swelling behaviour by nitrile rubber compositions is generally assigned to the "like dissolves like" principle because biodiesel is an ester and it consequently possesses some polarity (HASEEB et al., 2010; HASEEB et al., 2011b; ALVES; MELLO; MEDEIROS, 2013; CHAI et al., 2013; ZHU et al., 2015). The closeness of the polarities of the fluid and elastomer would allow them to interact, facilitating the diffusion of the fuel into the polymer. However, given the obtained results, this explanation seems insufficient to fully describe the differences in the resistance that the nitrile rubber compositions exhibited to biodiesel.

Additionally, Haseeb et al. (2010; 2011b) proposed that an increase in the acrylonitrile content of a composition would increase its crosslink density, which would consequently reduce the biodiesel swelling process. However, the authors did not make efforts to support their conjectures. Based on the results herein presented, these conjectures cannot be fully accepted.

No relation between crosslink densities and oil mass uptake levels was observed (Figure 15). Both non-carboxylated nitrile rubber compositions (*NBR33* and *NBR45*) were formulated using the same curing system, and yielded to similar crosslink densities as discussed in section 5.1.1.2 and shown in Figure 13; however, these compositions swelled to different extents.

Furthermore, *XNBR*, which possessed the highest crosslink density, exhibited behaviour between those of the two other compositions, which could have been due to the different types of crosslinks formed. An additional type of crosslink network, i.e., ionic crosslinks, may have compensated for the lower acrylonitrile content of the sample.

It can be suggested that solely changing the crosslink density has a small effect on the interactions between nitrile rubber samples and biodiesel. Rather, the acrylonitrile content and the type of crosslink networks play more significant roles in the interactions between fuels and elastomers.

Figure 15 – Changes in mass after 22 h of immersion in soybean biodiesel as a function of crosslink densities of the nitrile rubber compositions.

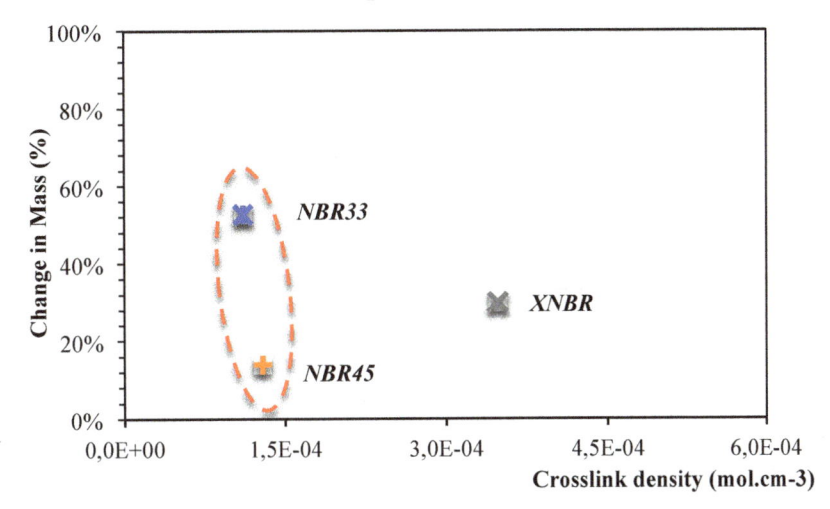

5.1.2.2 Physical mechanical resistance

The relative changes observed through the physical mechanical tests are presented in Figure 16, along with the absolute values before and after the samples were immersed in soybean biodiesel.

Overall, all the nitrile rubber samples exhibited poor performances. As expected, based on the changes observed in the mass tests, the *NBR33* samples presented the worse results; this composition lost, on average, 66% of the its initial properties. Remarkably, the tear strength test showed that this composition lost more than 80% of its initial value. The elongation at break test, likewise, showed that the composition lost approximately 70% of its initial resistance.

Nitrile rubber that contains 33% acrylonitrile is usually the most commonly employed type of *NBR* in the production of most nitrile rubber-based articles. Based on the results presented, the use of this NBR should be highly discouraged in applications in which pure biodiesel is present.

Although *NBR45* and *XNBR* showed slightly better resistance to biodiesel, their mechanical losses were still significant. They lost on average 38% and 41% of their initial properties, respectively. The tensile strengths for both compositions presented similar losses (52% and 48%, respectively) and similar absolute values after immersion (9.5 MPa and 10.5 MPa, respectively).

Figure 16 – Physical mechanical test results of non-immersed (dark colours) and immersed (light colours) nitrile rubber compositions. Between brackets indicate the percentage of loss of each composition.

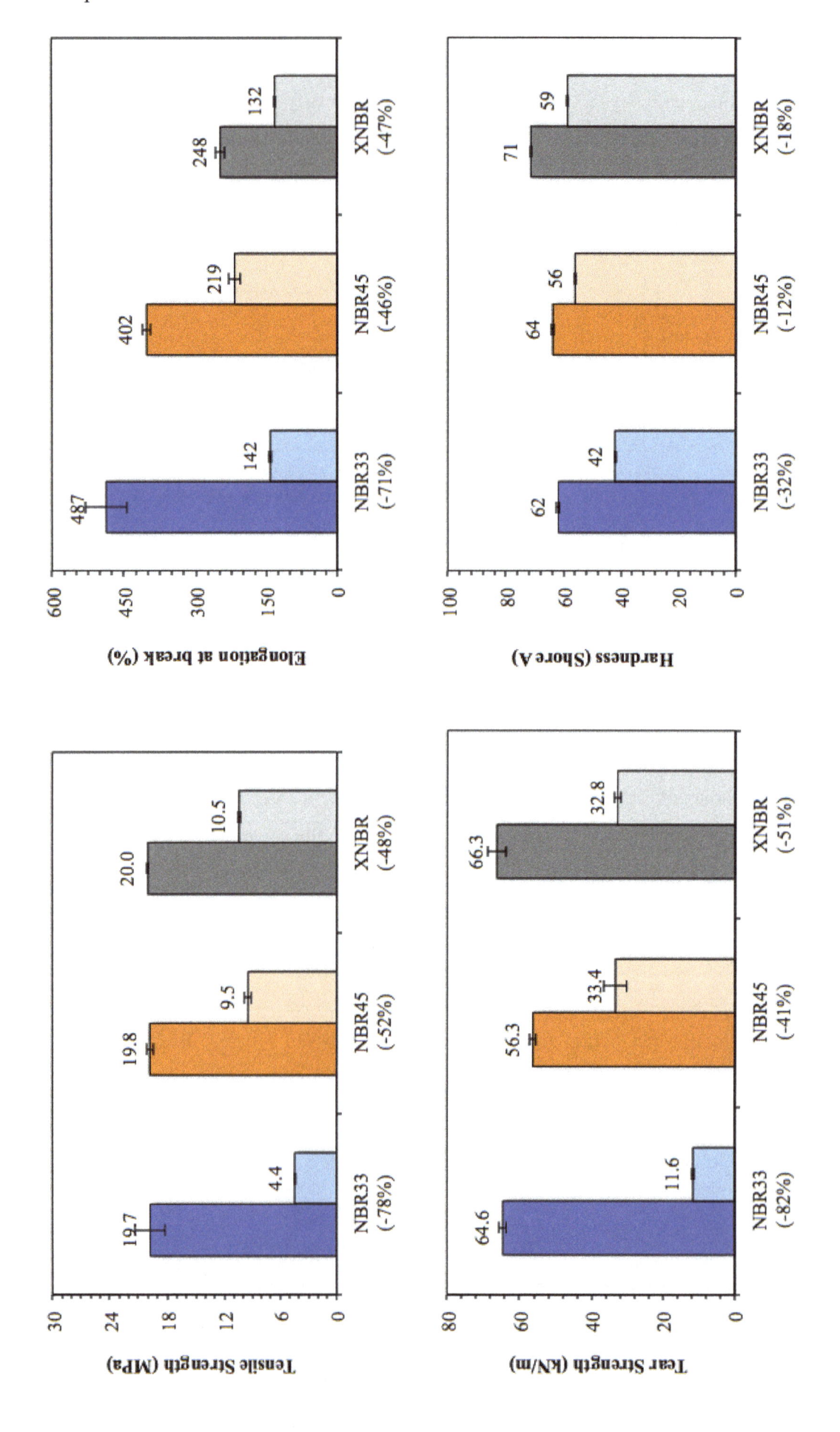

These results followed trends that were already observed in previous studies (LINHARES; FURTADO, 2008; LINHARES et al., 2013), which showed increases in resistance to biodiesel with increases in acrylonitrile content, in spite of the different temperatures employed during the tests.

The mechanical losses observed could be attributed to a series of physical and chemical interactions between biodiesel and the compositions: a reduction in the polymer chains entanglement (TRAKARNPRUK; PORNTANGJITLIKIT, 2008), oxidation of the elastomer matrix-free double bonds, reactions with the crosslink systems, and reduction of the polymer-filler interactions (AKHLAGHI et al., 2015b).

Biodiesel absorption may act as a plasticiser, easing the mobility of the polymer chains and reducing the chain entanglement (TRAKARNPRUK; PORNTANGJITLIKIT, 2008). This plasticisation effect led to a substantial decrease in the observed mechanical properties, especially regarding elongation at break.

The detrimental effects of the oil also come from its low oxidative stability (GIAKOUMIS, 2013; SERRANO et al., 2013; JAKERIA; FAZAL; HASEEB, 2014; SANGEETA et al., 2014; BERGTHORSON; THOMSON, 2015), which is due to the high presence of unsaturated components (GIAKOUMIS, 2013; SANTOS et al., 2013; SERQUEIRA et al., 2014). The oxidation products may have reacted with the rubber (AKHLAGHI et al., 2015b), causing degradation, which also reduced the mechanical properties.

Hardness changes after immersion of *NBR45* and *XNBR* samples were only slightly affected after the immersion process (-12% and -18%, respectively). These results are in agreement with the mass-change analyses, which indicated that the samples were less swollen than *NBR33*. In addition, one can conclude that these samples were likely more affected by physical interactions with biodiesel than by chemical degradation. Moreover, the hardness of *NBR33* was more seriously affected after the immersion process, which suggested that chemical interactions may have occurred between this sample and biodiesel, i.e., the degradation of the crosslink network (AKHLAGHI et al., 2015b).

The losses in tensile strength after immersion were matched with each composition's crosslink densities (Figure 17). Once again, these changes showed that crosslink density do not solely rule rubber's resistance to biodiesel. *NBR33* and *NBR45*, which have similar crosslink densities, presented remarkably different mechanical resistance to biodiesel, which may be assigned to differences in the acrylonitrile content.

Moreover, *XNBR*, which had the highest crosslink density, exhibited a similar loss to *NBR45*. In contrast to what was proposed by Haseeb et al. (2010; 2011b), an increase in crosslink density did not directly imply improved resistance of a rubber to biodiesel. One can infer that different types of crosslink networks compensate for the lower acrylonitrile content. This analysis agreed well with observations on the mass uptake as a function of the crosslink densities of the compositions (Figure 15).

Figure 17 – Tensile strength change (%) after 22 h of immersion in soybean biodiesel at 100°C as a function of crosslink density of the nitrile rubber composition.

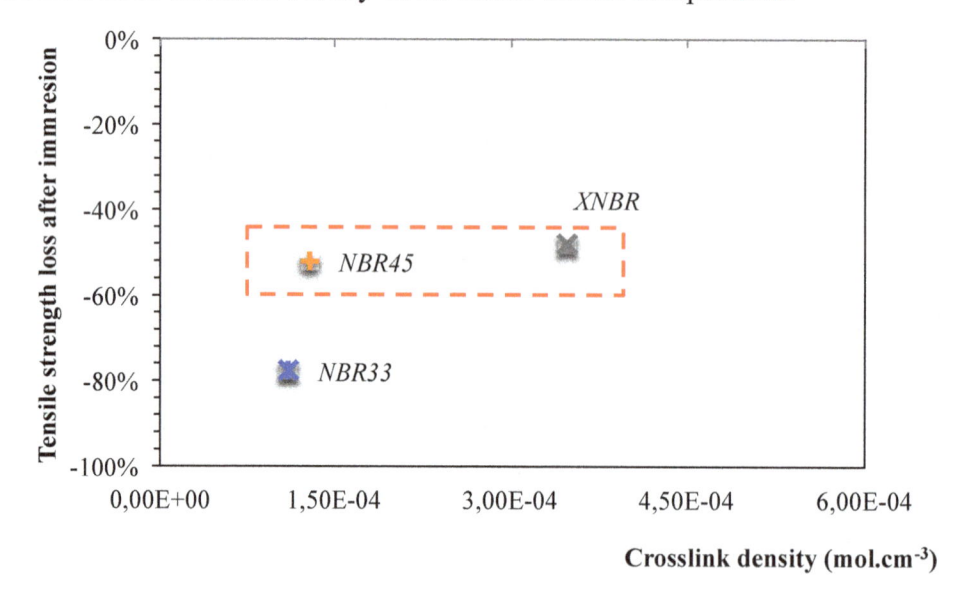

The losses of tensile strength were also compared to the mass uptake of each composition (Figure 18). Interestingly, the mechanical losses were not directly affected by oil uptake. Indeed, *NBR33* absorbed the greatest amount of biodiesel and lost its mechanical properties in the highest extent. However, *XNBR* absorbed more biodiesel than *NBR45*, but both presented similar losses in tensile strength. This suggests that plasticisation and further detrimental effects of biodiesel on *XNBR* were less prominent compared to the other compositions.

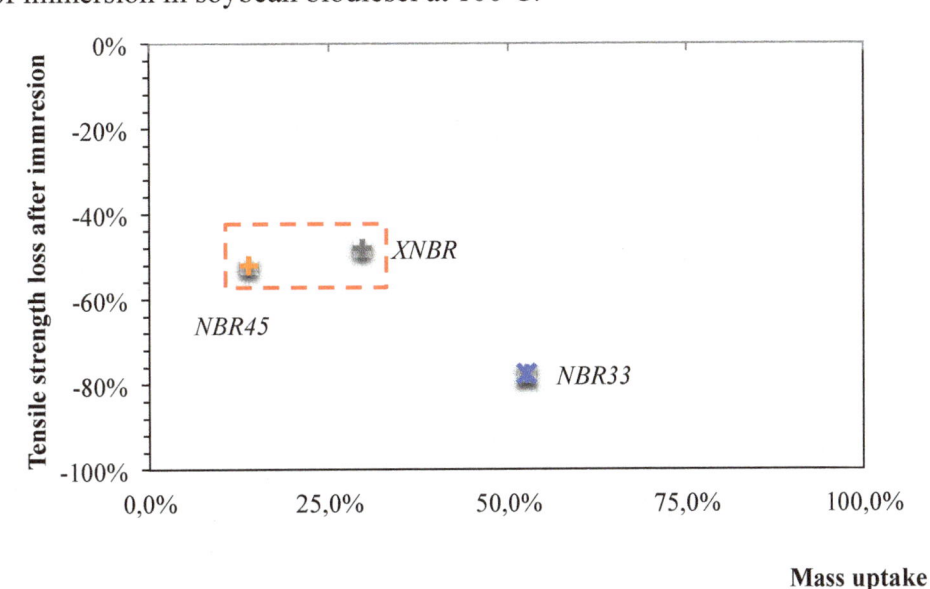

Figure 18 – Tensile strength change (%) as a function of the mass uptake of each composition after 22h of immersion in soybean biodiesel at 100°C.

5.1.2.3 Scanning Electron Microscopy (SEM)

Photomicrographs of the fracture surfaces of the compositions are displayed in Figure 19. Before immersion, no voids, cracks or clusters were observed on the surfaces of the compositions.

Moreover, upon immersion in biodiesel the surface of the *NBR33* composition was observed to be modified and highly deteriorated compared to that of the non-immersed sample. Many clusters were formed on the surface of the immersed sample, whose presence indicated the occurrence of a chemical interaction between biodiesel and the elastomer matrix or a reaction between biodiesel and some of the formulation ingredients. The modified surface indicated a strong, but destructive, interaction between biodiesel and the elastomer.

Observations of degraded surfaces of nitrile rubber compositions (i.e., pits and cracks) were also reported in the literature (HASEEB et al., 2010; ALVES; MELLO; MEDEIROS, 2013; CORONADO et al., 2014). However, these studies failed to provide information on which nitrile rubber grade was employed in their experiments, which avoided to proper compare the results. Akhlaghi et al. (2015b) utilised SEM and found that the rubber layers covering carbon black particles were removed after immersion in biodiesel; however their tests were conducted with an even lower acrylonitrile-content nitrile rubber sample (28%).

The obtained photomicrographs corroborate the suggestion that some chemical interactions occur between the oil and this composition.

Figure 19 – Scanning electron microscopy (SEM) photomicrographs of nitrile rubber compositions: non-immersed and after immersion for 22 h at 100°C in soybean biodiesel.

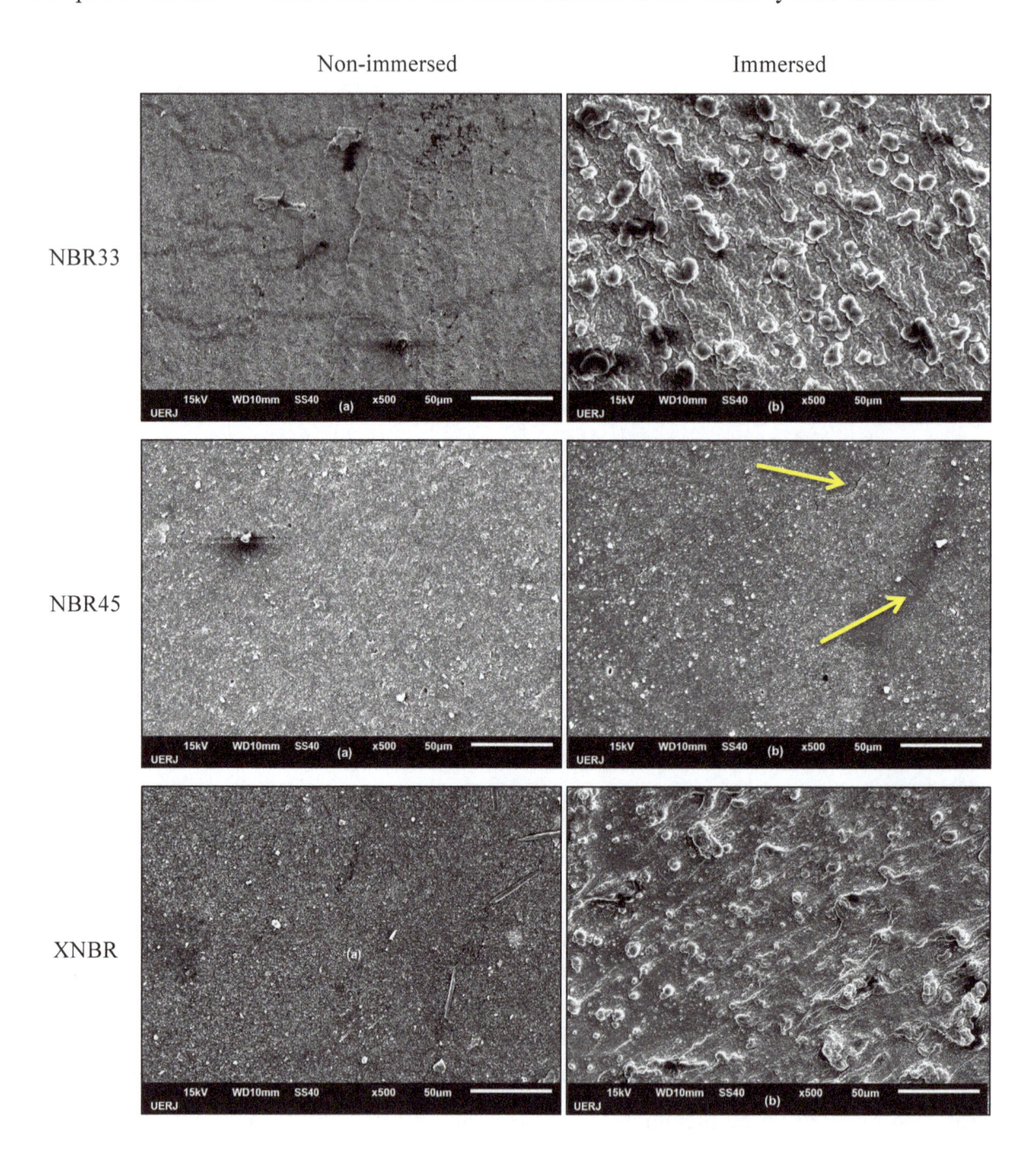

	Non-immersed	Immersed
NBR33		
NBR45		
XNBR		

The surface of the *NBR45* composition showed little signs of attack after the immersion process, as almost no modifications were observed by SEM analysis. Only a few discreet cracks could be noted on the surface of the composition. The obtained photomicrographs are in agreement with the other tests, as this composition was shown to be more resistant than the others. Previous work (LINHARES et al., 2013) also showed less modification of the surface with an increase in the acrylonitrile content in the rubber.

The carboxylated nitrile rubber composition exhibited fewer clusters than the *NBR33* composition after immersion in biodiesel, despite its lower acrylonitrile content. However, the surface was more degraded than that of *NBR45*. Although the mechanical behaviour of the *XNBR* was similar to the high acrylonitrile content composition (*NBR45*), the photomicrographs suggested that some chemical interactions between *XNBR* and the biodiesel had occurred.

5.1.3 Overall performance

The compositions *NBR45* and *XNBR* showed the best mechanical results after immersion in biodiesel among the samples that were tested. Despite having a low acrylonitrile content (28%), the modified nitrile rubber samples exhibited a higher resistance to biodiesel compared to the medium acrylonitrile rubber content (*NBR33*). As already mentioned, *XNBR* presented an average loss close to that of the high-acrylonitrile-content samples (*NBR45*). Both the acrylonitrile content and different types of crosslink network increased the samples' mechanical resistance to biodiesel.

Based on the mechanical results and morphological analyses, the composition *NBR45* was chosen as a base to continue the study on improving nitrile rubber resistance to biodiesel. Modifications of the sulphur-based curing system were proposed by changing the types and the amounts of accelerators employed in the formulation. The following assessment was conducted exclusively with a nitrile rubber sample containing 45% of acrylonitrile.

5.2 Part II – Formulation development: The influence of binary sulphur-based curing systems

5.2.1 Characterisation of the compositions

5.2.1.1 Crosslink density and differential scanning calorimetry (DSC)

The compositions prepared in Part II were labelled according to the amount in phr (part per a hundred parts of rubber) of each vulcanisation component employed in each formulation, following the order TMTD/CBS/sulphur.

Considering the ratio between the amounts of accelerators and sulphur employed in the formulations and the factors that were described in Table 4 (section 1.4.2), the prepared compositions were divided according to their vulcanisations systems (Table 12). Composition *2(1/0/1.5)* utilised a conventional vulcanisation (CV) system, i.e., it would mainly contain polysulphide crosslinks, whereas compositions *3(1/2/0.5), 5(3/0/0.5), 7(3/2/0.5), 8(3/2/1.5),* and *9(2/1/1)* utilised an efficient vulcanisation (EV) system, i.e., they would mainly contain monosulphide crosslinks. The other compositions used a semi-efficient vulcanisation (SEV) system.

Table 12 – Vulcanisation systems of each prepared nitirle rubber composition. Numbers between brackets indicate the amount in phr of each curing system component of the formulations (TMTD/CBS/sulphur).

Composition	Accelerator/sulphur ratio	Vulcanisation system
1(1/0/0.5)	2	SEV[a]
2(1/0/1.5)	0.7	CV[b]
3(1/2/0.5)	6	EV[c]
4(1/2/1.5)	2	SEV
5(3/0/0.5)	6	EV
6(3/0/1.5)	2	SEV
7(3/2/0.5)	10	EV
8(3/2/1.5)	3.3	EV
9-A(2/1/1)	3	EV
9-B(2/1/1)	3	EV
9-C(2/1/1)	3	EV

Footnote: (a) – semi-efficient vulcanisation system; (b) – conventional vulcanisation system; (c) – efficient vulcanisation system

According to Datta et al. (2007), combining the accelerators TMTD and CBS would lead to the formation of higher amount of monosulphide crosslinks, regardless of the ratio employed. Based on this hypothesis, most of the prepared compositions would then also be considered to have EV systems. Nevertheless, these compositions were treated as SEV system compositions throughout this Thesis, supported by the considerations exposed in Table 4 (section 1.4.2).

The crosslink densities of the compositions are shown in Figure 20. The highest crosslink density was achieved by composition *8(3/2/1.5)*, which was formulated with the highest amount of each of the curing system components. Analysis of variance (ANOVA) showed that only sulphur (p=0,000485) and TMTD (p=0,00899) had an influence on the crosslink densities of the compositions. Given that TMTD is a sulphur-donor accelerator (LAWANDY; HALIM, 2005; MOVAHED; ANSARIFAR; MIRZAIE, 2015), this behaviour was expected.

In contrast to reports by some authors (DEBNATH; BASU, 1996; LAWANDY; HALIM, 2005; MOVAHED; ANSARIFAR; MIRZAIE, 2015), who indicated that CBS would decrease the crosslink density, CBS or the interaction between CBS and other curing system components did not have any statistical influence (p>0.05) on the crosslink densities considering the amounts of reagents that were employed.

Figure 20 – Crosslink densities (bars) and glass transition temperatures (line) of the nitrile rubber compositions. Numbers between brackets indicate the amount in phr of each curing system component of the formulations (TMTD/CBS/sulphur).

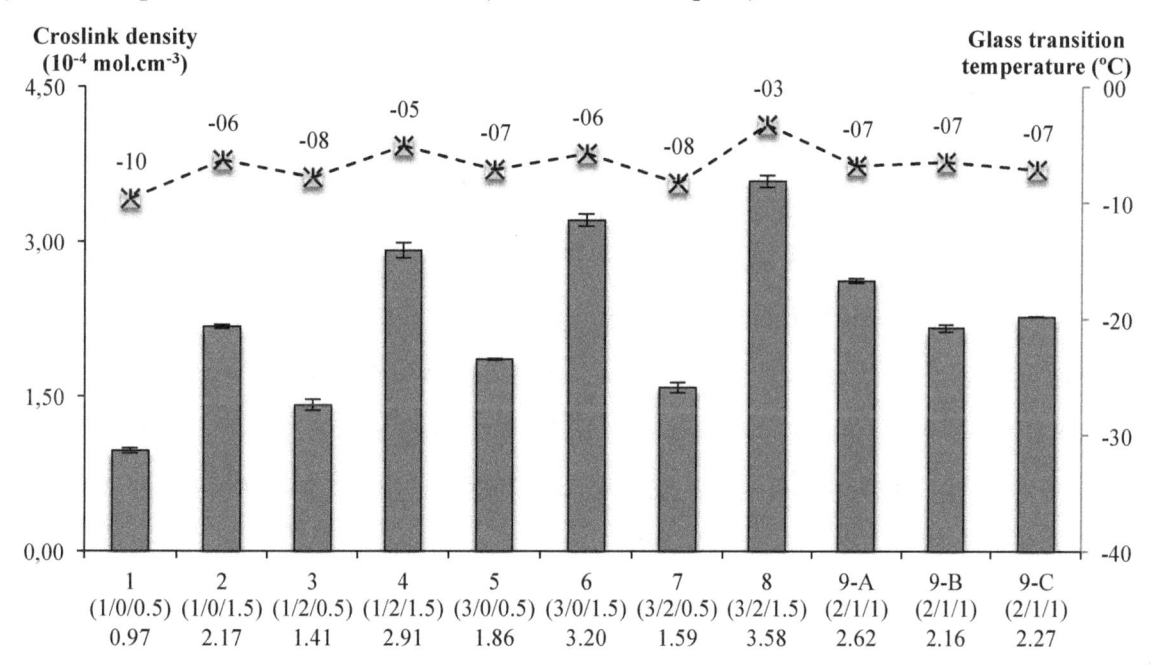

The glass transition temperatures (T_g) of the compositions are also shown in Figure 20. Both the crosslink density and the T_g followed the same trend: an increase in crosslinking of the rubber matrix required the amorphous polymeric phase to have more energy gain mobility (MANO; MENDES, 1999; ODIAN, 2004); therefore, the glass transition occurred at a higher temperature. Composition *8(3/2/1.5)* presented the highest T_g (-3.3°C), which was in line with the expected behaviour of the materials.

5.2.1.2 Mechanical properties

The mechanical results of the compositions are shown in Figure 21. The tensile strengths of the all compositions ranged between 19.1 MPa to 22.8 MPa, among which, composition *3(1/2/0.5)* exhibited the highest tensile strength. Based on the ANOVA results, tensile strength was statistically affected by the amount of TMTD (p=0.010008), the amount of sulphur (p=0.002494), and the interaction between TMTD and sulphur (p=0.011505).

No relations between the vulcanisation systems and tensile strengths were noted. Both of these properties are believed to be connected (NASIR; G. K. TEH, 1988; GONZÁLEZ et al., 2005a; GONZÁLEZ et al., 2005b; LAWANDY; HALIM, 2005), regardless of the crosslink density. Nevertheless this seems to be true only for crystallising rubbers and it is not valid for NBR.

The differences in elongation at break (%) of the compositions were more pronounced than the observations of tensile strength, and the values ranged from 196% to 508%. Additionally, ANOVA indicated that both TMTD (p=0.004407) and sulphur (p=0.000082) statistically influenced elongation at break properties. Moreover, elongation was also affected by the interaction between TMTD and CBS (p=0.041141).

Hardness was affected by the amount of sulphur (p=0.00024), the amount of TMTD (p=0.0142) and the interaction between CBS and sulphur (p=0.022365) in the formulation, as indicated by ANOVA. Similar to the observations obtained on the effects of crosslink density, the highest value of hardness was achieved by the composition that was formulated with the highest amount of each of the curing system components. However, the differences among the compositions were not very pronounced.

The Pareto charts of the standardised effects of the variables and the response surfaces are depicted in APPENDIX D in Figure 53, and Figure 54.

Figure 21 – Mechanical test results of the nitrile rubber compositions. Between brackets indicate the amount in phr of each curing system component of the formulations (TMTD/CBS/sulphur).

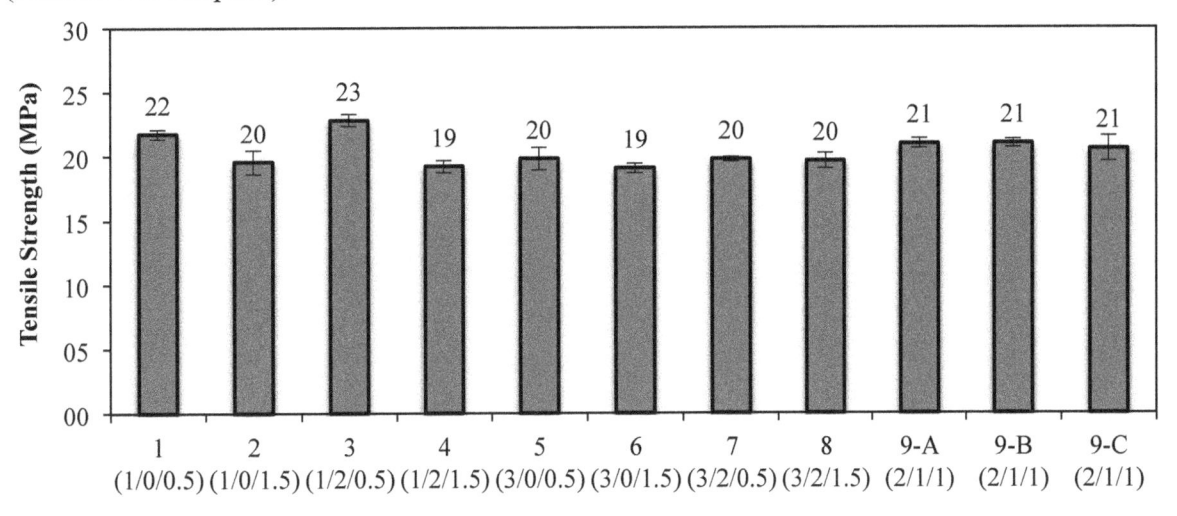

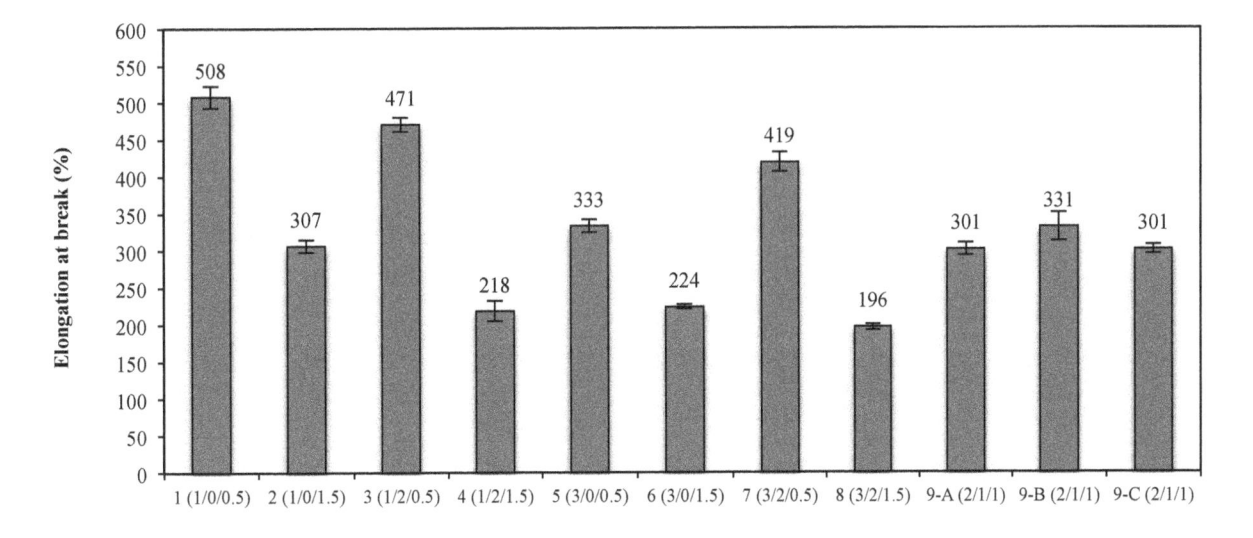

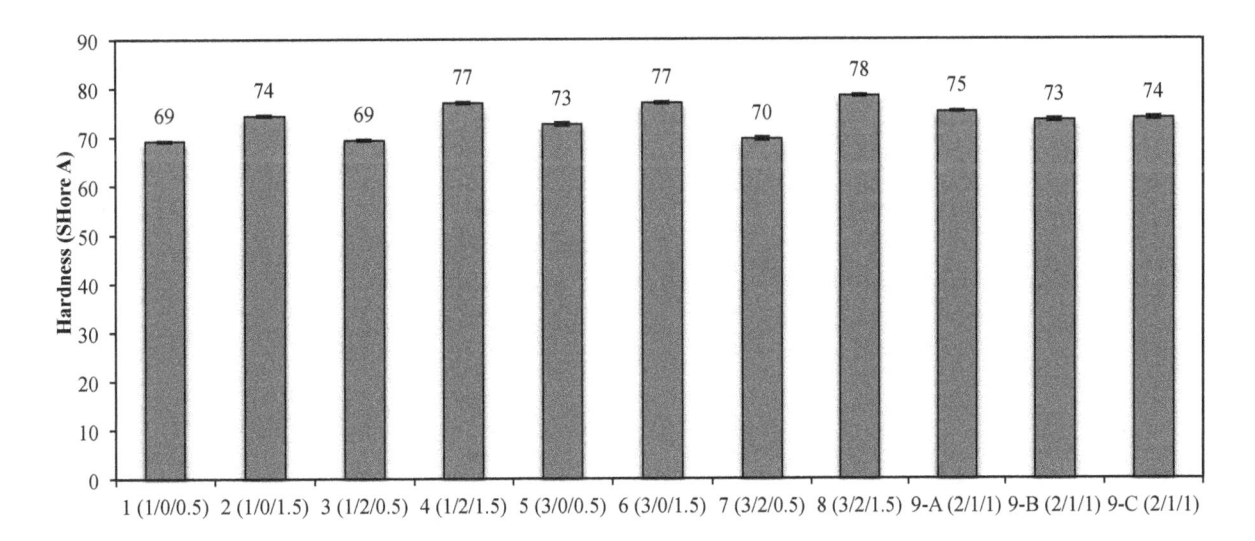

The effect of crosslink density on tensile strength was assessed considering the presence and absence of CBS (Figure 22). This approach was necessary because CBS can modify the vulcanisation mechanism.

Figure 22 – Tensile stregths (MPa) of the nitrile rubber compositions as a function of their crosslink densities. Numers between brackets indicate the amount in phr of each curing system component of the formulations (TMTD/CBS/sulphur).

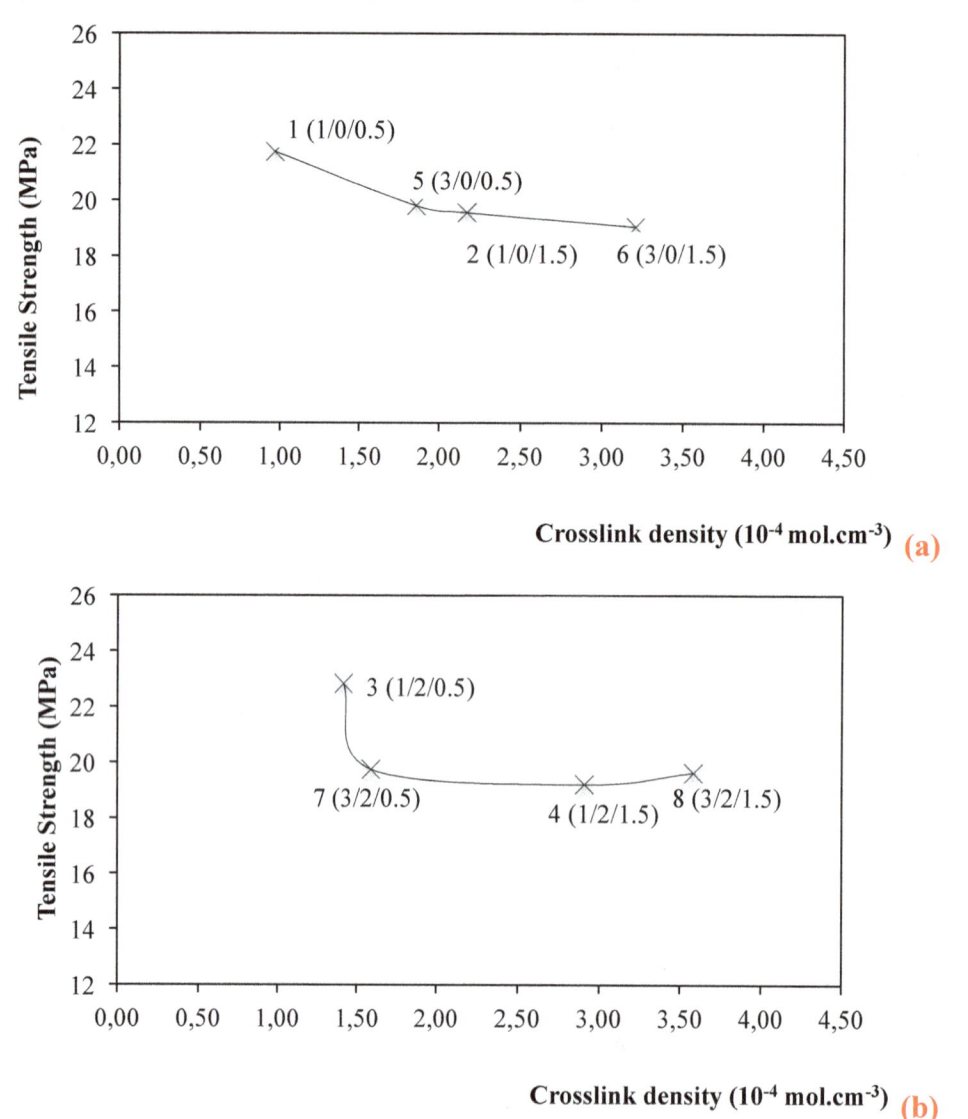

(a) Tensile Strength (MPa) vs Crosslink density (10^{-4} mol.cm^{-3})
- 1 (1/0/0.5)
- 5 (3/0/0.5)
- 2 (1/0/1.5)
- 6 (3/0/1.5)

(b) Tensile Strength (MPa) vs Crosslink density (10^{-4} mol.cm^{-3})
- 3 (1/2/0.5)
- 7 (3/2/0.5)
- 4 (1/2/1.5)
- 8 (3/2/1.5)

Footnote: (a) – Compositions **without** the accelerator CBS on the formulation;
(b) – Compositions **with** the accelerator CBS in the formulation.

As a rule, tensile strength does not have a linear relation with the crosslink density of a material; it passes through a maximum point, after which the tensile strength decreases (CORAN, 2003; 2005; GONZÁLEZ et al., 2005a; GONZÁLEZ et al., 2005b; EL-NEMR, 2011).

Regardless of the criteria employed, the tensile strength decreased with an increase in crosslink density. Hence it is believed that the prepared compositions were beyond the crosslink density optimum point for tensile strength.

Elongation at break decreased linearly with an increase in crosslink density, whereas hardness increased almost linearly (Figure 23). These finding were in complete agreement with findings reported in the literature (CORAN, 2003; 2005).

The results obtained through these analyses agree well with one another, including observations related to increases in crosslink density and the way the materials became harder and more brittle, which reduced the tensile strength and the rubber elongation capacity.

Figure 23 – Elongation at break and hardness as a function of crosslink density. Numbers between brackets indicate the amount in phr of each curing system component of the formulations (TMTD/CBS/sulphur).

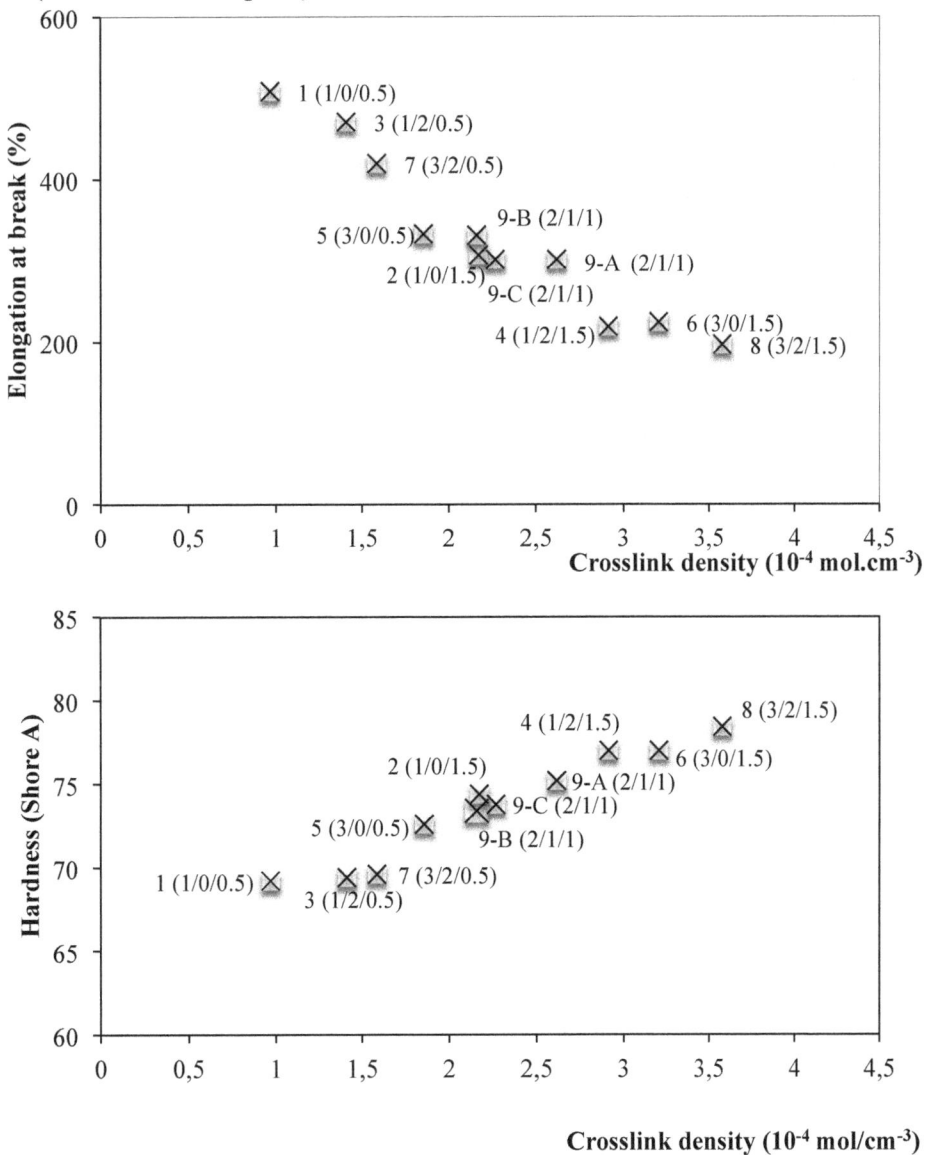

5.2.1.3 Dynamic mechanical thermal analysis (DMTA)

The loss and elastic moduli of the nitrile rubber compositions are presented in Figure 24, and the glass transition temperature (T_g) of the compositions obtained at the temperature of the tan δ maxima are shown in Table 13.

Figure 24 – Elastic (a) and viscous (b) moduli *versus* temperature of the nitrile rubber compositions obtained at 1Hz. Numbers between brackets indicate the amount in phr of each curing system component of the formulations (TMTD/CBS/sulphur).

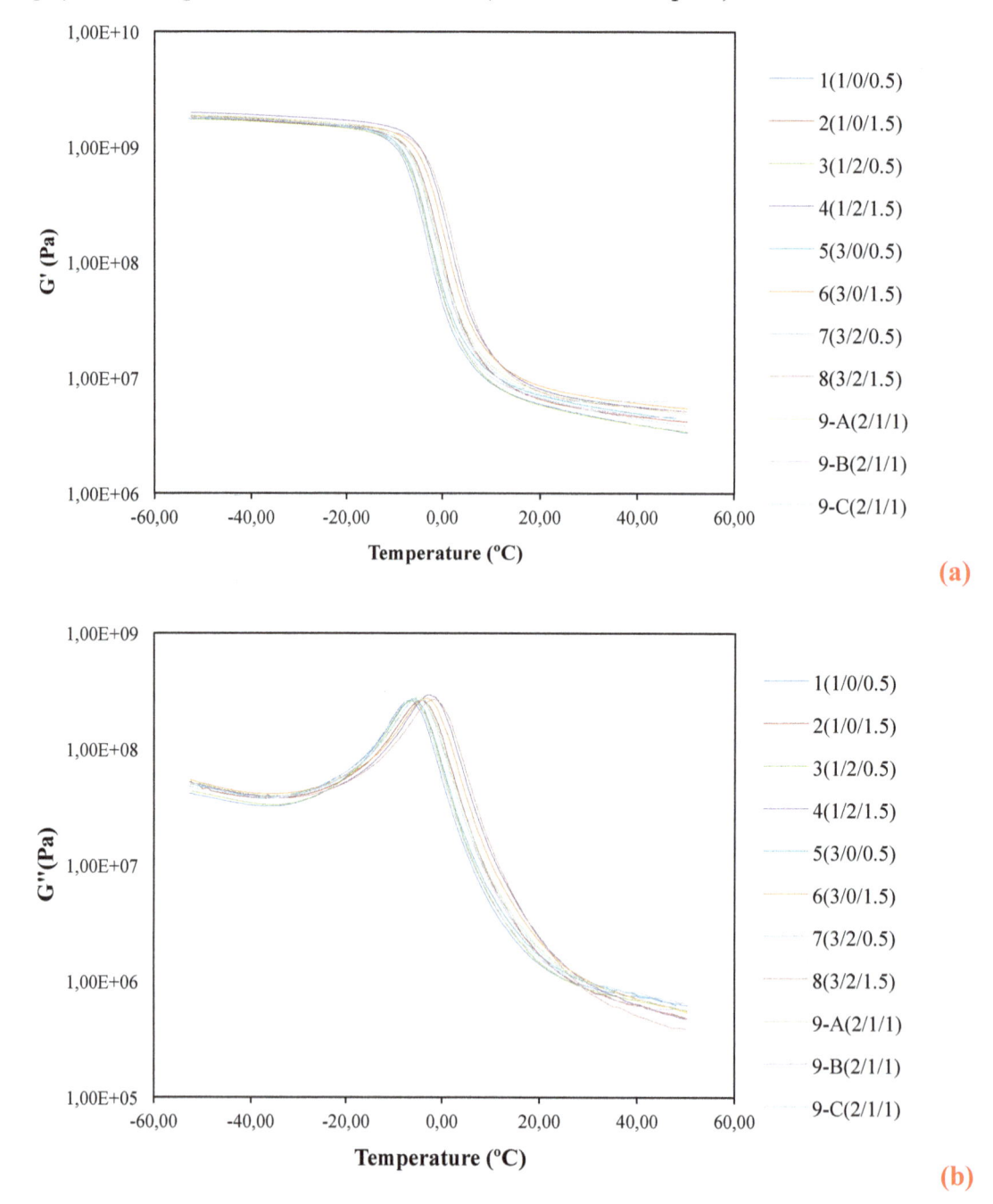

It could be observed that the moduli of the compositions did not present significant differences among them, either at low or high temperatures (Figure 24). The only observed changes were with respect to the T_g (Table 13) when they were obtained as the temperatures of the maximum tan δ peaks (LIU et al., 2014; SPRENGER; KOTHMANN; ALTSTAEDT, 2014). The different T_g value followed the same trend that was observed by the DSC analyses. However, the temperatures given by DMTA were slightly higher than those given by DSC (RAHMAN; AL-MARHUBI; AL-MAHROUQI, 2007; JOGI et al., 2014; LIU et al., 2014), as expected, given the different nature of the two methods.

Table 13 – Glass transition tramperature (°C) of the compositions obtained from the tan maxima. Numbers between brackets indicate the amount in phr of each curing system component of the formulations (TMTD/CBS/sulphur).

Composition	Glass transition temperature (°C)
1(1/0/0.5)	-0,13
2(1/0/1.5)	1,98
3(1/2/0.5)	0,68
4(1/2/1.5)	4,49
5(3/0/0.5)	0,45
6(3/0/1.5)	2,87
7(3/2/0.5)	-0,04
8(3/2/1.5)	4,84
9-A(2/1/1)	1,45
9-B(2/1/1)	1,86
9-C(2/1/1)	1,23

5.2.1.4 Scanning electron microscopy (SEM)

Scanning electron microscopy (SEM) photomicrographs of four different compositions were chosen, based on their TMTD content to exemplify the prepared compositions. Low TMTD compositions *3(1/2/0.5)* and *4(1/2/1.5)* presented no special features on their fracture surfaces (Figure 25) and were free of voids, cracks, or agglomerates. Additionally, the filler seemed well dispersed and distributed.

Compositions *5(3/0/0.5)* and *7(3/2/0.5)* were chosen to represent the high TMTD-content compositions. Their photomicrographs revealed the presence of some crystals on the fracture surfaces of the high TMTD-content compositions. These crystals were also observed in the other high TMTD-content compositions.

Figure 25 – SEM fracture surface photomicrographs of the nitrile rubber compositions. Numbers between brackets indicate the amount in phr of each curing system component of the formulations (TMTD/CBS/sulphur).

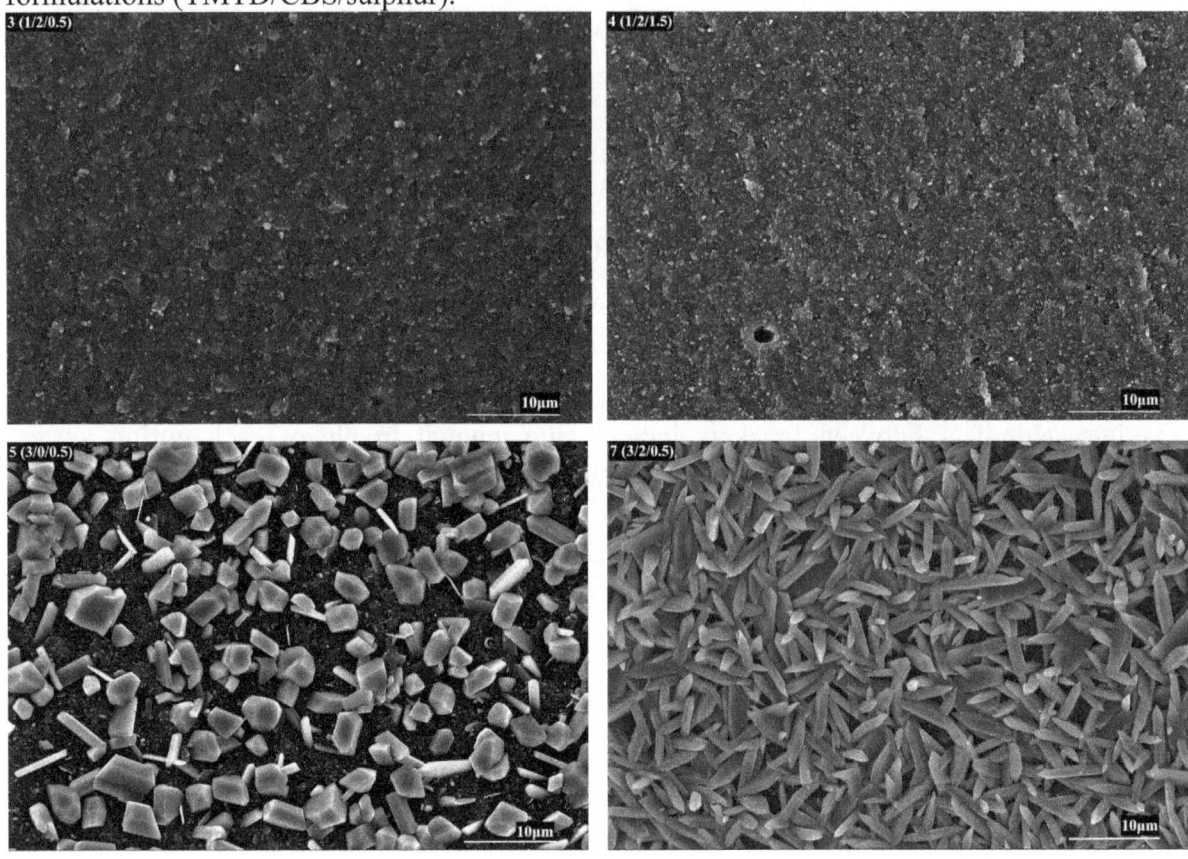

The photomicrographs of the non-fractured surface of compositions *3(1/2/0.5)* and *7(3/2/0.5)* are depicted in Figure 26. Additionally, the presence of some crystals on the surface of the high TMTD-content compositions was noted. These crystals may be related to the blooming of some of the curing system components, which has been reported when high amounts of TMTD are employed in rubber formulations (FRANTA, 1989).

Figure 26 – SEM non-fractured surface photomicrographs of the nitrile rubber compositions. Numbers between brackets indicate the amount in phr of each curing system component of the formulations (TMTD/CBS/sulphur).

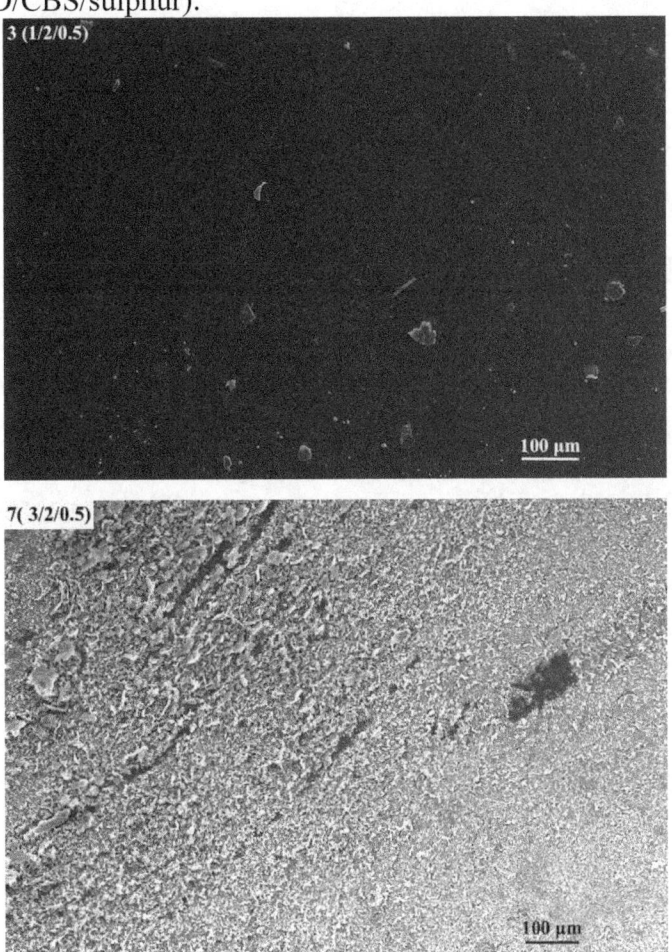

A cryogenic break test was conducted with composition *7(3/2/0.5)* to better understand the tension and/or time dependence of the observed exudation (Figure 27). The cryogenic-fracture surface photomicrograph of the part of the composition, which had been recently fractured, did not show any crystals or exuded components, in contrast, the part of the composition which rested for 80 days prior to the analysis showed crystals on the surface.

Photomicrographs of the formulation's pure components were also obtained and are presented in Figure 46 in APPENDIX B.

Based on these photomicrographs, one can infer that blooming is related to the high content of TMTD in the compositions; in addition, this phenomenon was found to be time dependent. One can also infer that the observed exudation was also tension dependent, owing to the fact that the tensioned samples presented higher amounts of the crystals on their surfaces than those that were manually fractured.

Figure 27 – SEM fracture surface photomicrographs after cryogenic fracture of nitrile rubber composition with a high TMTD content. Composition *7(3/0/0.5)*.

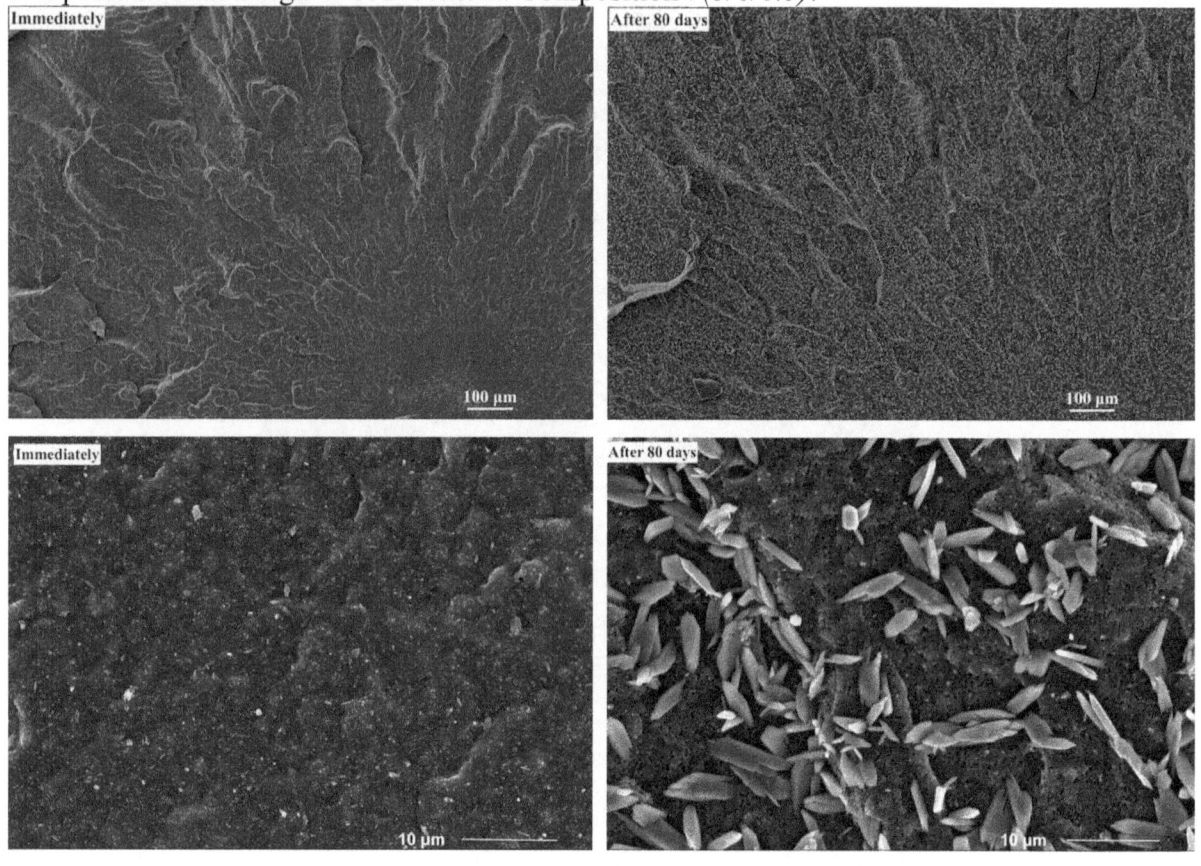

5.2.1.5 Confocal Laser Scanning Microscopy (CLSM)

The fracture surfaces of the compositions were also analysed by Confocal Laser Scanning Microscopy (CLSM), and they are shown in Figure 28. The same compositions chosen for SEM analyses were also used for CLSM analyses to represent both low and high TMTD-content compositions. These photomicrographs corroborated the analyses conducted by SEM. Low TMTD-content compositions exhibited smooth surfaces that were free of voids, agglomerates or cracks, whereas crystals could be observed on high TMTD-content fracture surfaces. This analysis backs the hypothesis that TMTD or its related components migrated (exuded) to the surface of the compositions.

Figure 28 – CLSM photomicrographs of the nitrile rubber compositions. Numbers between brackets indicate the amount in phr of each curing system component of the formulations (TMTD/CBS/sulphur).

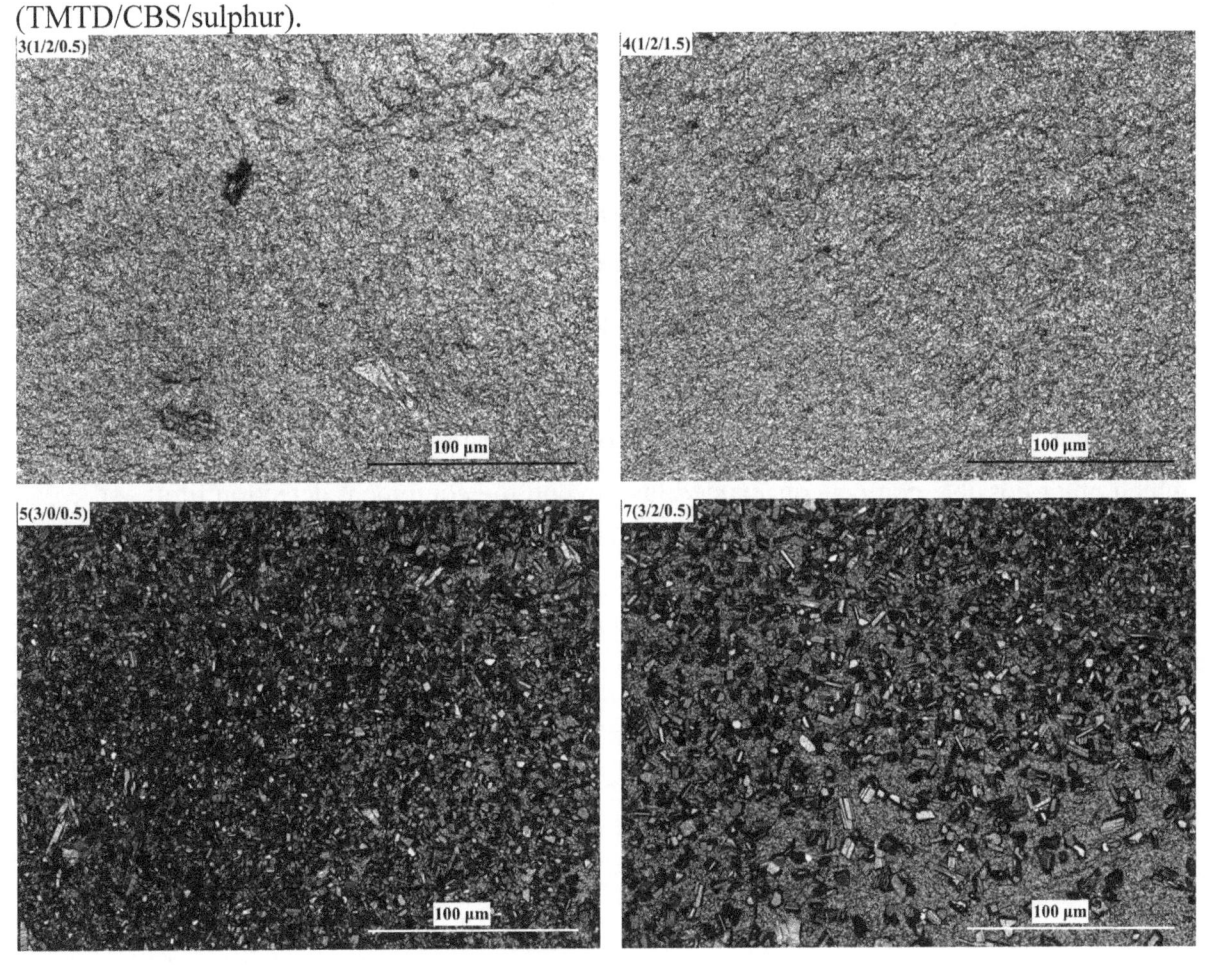

5.2.1.6 Attenuated total reflectance Fourier transform infrared (ATR-FTIR) spectroscopy

Compositions *3(1/2/0.5)* and *7(3/2/0.5)* were chosen as representatives of low TMTD-content compositions and for high TMTD-content compositions, respectively. The spectra of the surface of both compositions are shown in Figure 29. The FTIR spectra of pure TMTD and pure CBS are shown in Figure 47 and Figure 48, which can be found in APPENDIX C.

Some bands could be observed in both spectra, which could be attributted to the nature of the nitrile rubber. Nevertheless, some bands, which are highlighted in the spectra, could only be observed for the composition that experienced visual exudation.

One of the main different bands was (1) at 1239 cm^{-1}, which was assigned to N-C=S (SILVERSTEIN; WEBSTER; KIEMLE, 2005). This band was also observed in the spectrum of pure TMTD at 1232 cm^{-1}. However, a band at 1235 cm^{-1} was observed in the pure CBS spectrum, which could be attributed to the N-H vibrations of amines. These bands may have overlapped with each other, which would generate false results in the attempt to identify which component had been exuded.

The other band observed only for the exuded composition spectrum was (2) at 1386 cm^{-1}, which was assigned to C-H from ~CH_3 (SILVERSTEIN; WEBSTER; KIEMLE, 2005; COATES, 2006). This band differed from other types of C-H vibrations, e.g., from ~CH_2 or from ~CH~. This band was also observed in the pure TMTD spectrum at 1370 cm^{-1}. Moreover, no bands in this region were observed on the pure CBS spectrum. Based on their chemical structures, only TMTD has ~CH_3 groups; thus, one can infer that the exuded component was in fact the accelerator TMTD or a product of its decomposition. The extent to which other possible components exuded lay beyond the equipment detection limit and could be considered insignificant.

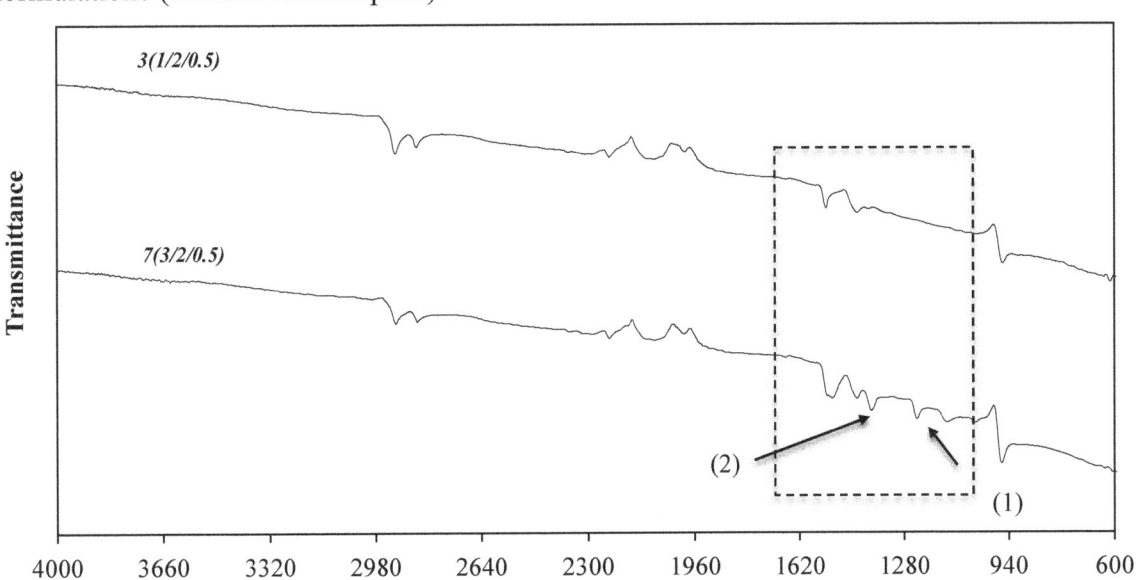

Figure 29 – FTIR spectra of the nitrile rubber compositions with low and high TMTD content. Numbers between brackets indicate the amount in phr of each curing system component of the formulations (TMTD/CBS/sulphur).

5.2.1.7 Nuclear magnetic resonance (NMR)

The NMR spectra of the solubilised components from the surfaces of the compositions 3(1/2/0.5) and 7(3/2/0.5), as well as the spectra of both the accelerators, TMTD and CBS, are shown in Figure 49 through Figure 52, which can be found in APPENDIX D.

No significant peaks were detected in the spectrum for composition *3(1/2/0.5)* (Figure 49), which confirmed that no components exuded to the surface of the composition. Nevertheless, significant signals at 3.484-3.494 ppm were observed in the spectrum of the composition *7(3/2/0.5)*, which experienced exudation (Figure 50). These signals were assigned to the protons of the methyl groups (SILVERSTEIN; WEBSTER; KIEMLE, 2005).

The spectrum of pure TMTD (Figure 51) presented a similar peak at 3.646-3.612 ppm, which was also assigned to methyl groups (BORDBAR; BIJANZADEH; ALIZADEH, 2004; SILVERSTEIN; WEBSTER; KIEMLE, 2005), whereas the spectrum of pure CBS (Figure 52) showed several different peaks: 1.228-1.295 ppm, 1.725-1.762 ppm, 1.773-1.781 ppm, and 2.080-2.100 ppm, which were assigned to the protons on its cyclic structure; 3.237-3.248 ppm, which was assigned to the N-H group; and 7.259-7.274 ppm, 7.372-7.387 ppm, 7.775-7.796 ppm, and 7.812 ppm, which were assigned to the aromatic structure (SILVERSTEIN; WEBSTER; KIEMLE, 2005).

Based on the fact that the exudation product possess peaks that can be assigned to methyl groups and only TMTD has methyl groups in its chemical structure (Table 15 in APPENDIX A), one can clearly infer that the exudation product is related to TMTD exudation, which is in agreement with results obtained from FTIR observations. Whether other components were exuding, it was below the equipment detection limit and can be considered insignificant.

5.2.2 Ageing tests

5.2.2.1 Gravimetric tests

The biodiesel uptake profiles after 700 h are depicted in Figure 30. During the first 166 h of the immersion test (shown in detail in Figure 30), only those compositions with EV systems presented apparent saturation in biodiesel, i.e., *(3(1/2/0.5), 5(3/0/0.5), 7(3/2/0.5), 9-A(2/1/1), 9-B(2/1/1),* and *9-C(2/1/1))*. This vulcanisation system yielded mono- and disulphides crosslinks, which are more thermally and chemically resistant (FRANTA, 1989; MANSILLA et al., 2015; MOVAHED; ANSARIFAR; MIRZAIE, 2015). It is expected that biodiesel can solubilise some additives of rubbers; however, during this test period, the absorption of biodiesel was greater than the loss of plasticisers, which was in agreement with the results reported by Akhlaghi et al. (2015b).

Interestingly, after 200 h of immersion, these compositions started showing again an increase in their weight, indicating they were no longer saturated. One can infer that after some initial resistance, the compositions started experiencing some actual degradation, which led to increasing in mass. According to Akhlaghi et al (2015b) and Haseeb et al. (2010; 2011b), biodiesel could have degraded the crosslink network of the nitrile rubber; in this sense, one can believe that the EV systems would start to deteriorate only after 200 h of immersion.

The other compositions, however, had a continuous increase in mass since the early stages of immersion until 500 h, after which they started to saturate. After the 700 h of the immersion test, the EV-compositions showed significantly lower change in mass than the other compositions.

Figure 30 – Mass change profiles of the nitrile rubber compositions after 700 h of immersion in soybean biodiesel at 100°C. Numbers between brackets indicate the amount in phr of each curing system component of the formulations (TMTD/CBS/sulphur). In detail, the mass change profiles after the first 166 h.

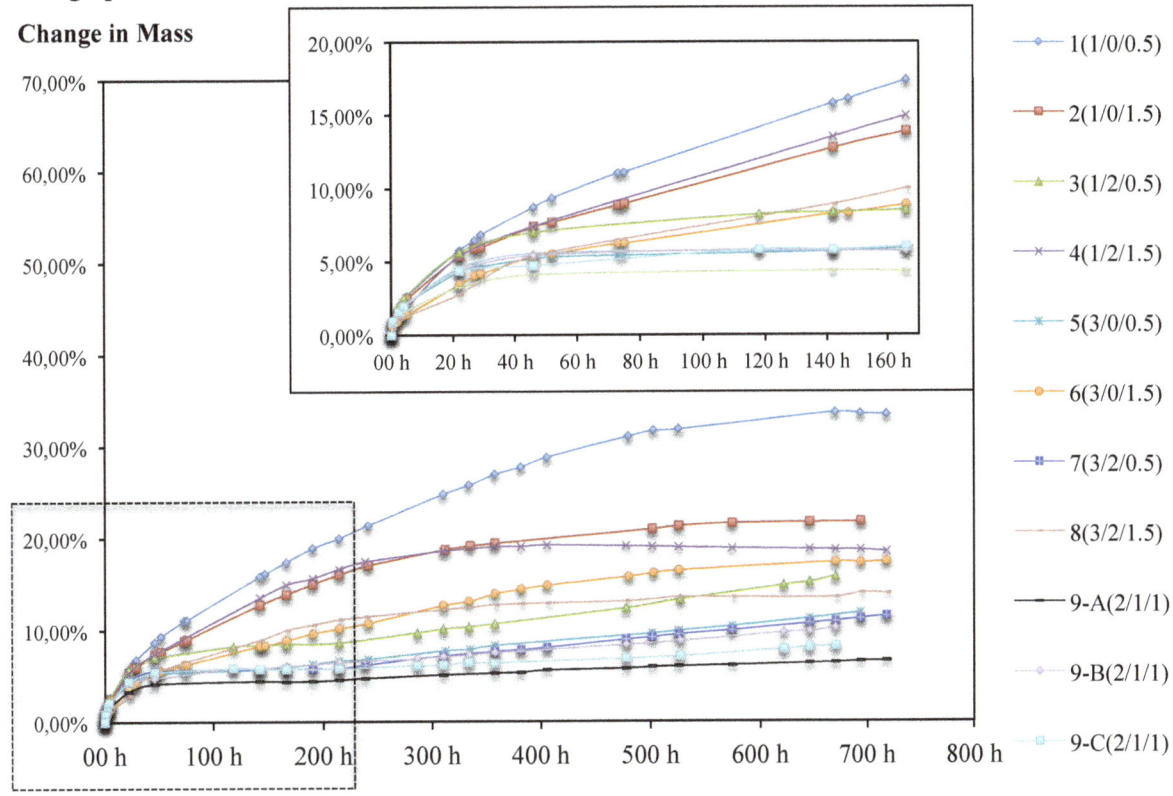

The change in mass after 22 h of immersion was only statistically affected by the amount of TMTD (Figure 55 in APPENDIX E). This can be understood by the fact that TMTD leads to more monosulphide crosslinks, which are more resistant. Nonetheless, the statistical influence on the mass uptake of a sample after 700 h would require the assessment of additional parameters.

Figure 31 relates the crosslink density and the change in mass of the compositions after 22 h. The low-TMTD-content compositions (circle markers) showed similar swelling behaviour regardless of their crosslink densities. Unlike for the high-TMTD-content compositions (cross markers), it could be concluded that the crosslink densities could play a role in the swelling process, i.e., they swelled less with an increase in crosslink density.

Figure 31 – Changes in mass after 22 h as a function of the crosslink densities of the nitrile rubber compositions. Numbers between brackets indicate the amount in phr of each curing system component of the formulations (TMTD/CBS/sulphur).

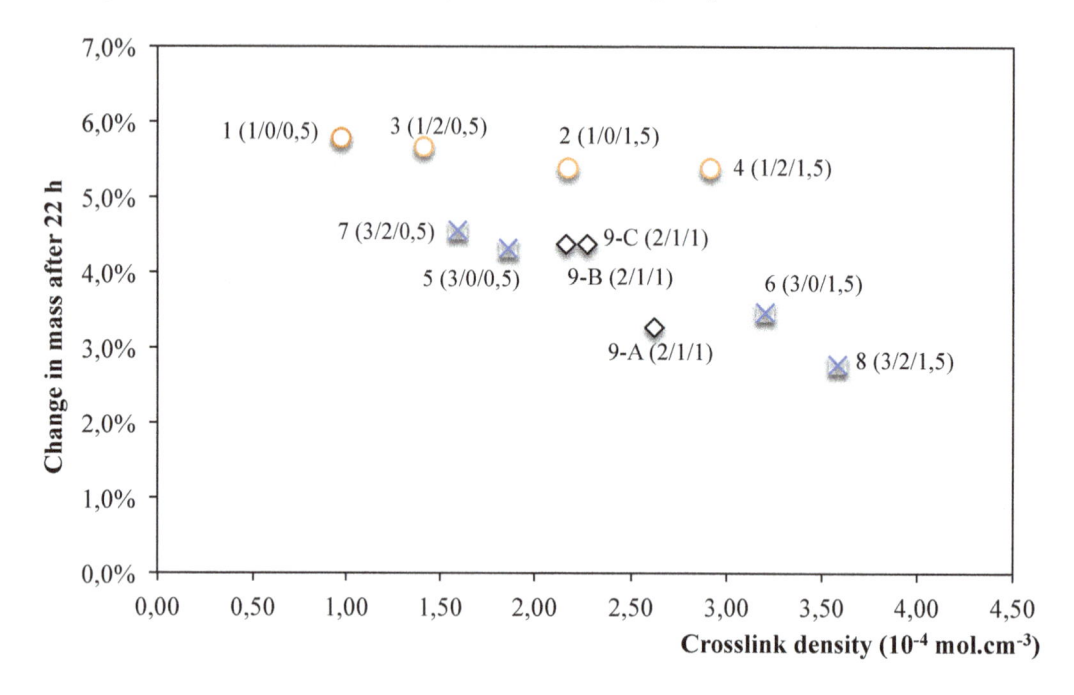

It could also be noted that compositions with similar crosslink densities swelled to significantly different extents during the first 22 h of immersion, e.g., composition *4(1/2/1.5)* swelled more than composition *6(3/0/1.5)*, despite their closeness in crosslink density, their same accelerator/sulphur ratios, and the consequent use of same vulcanisation system (SEV). Additionally, one can infer that the crosslink density does not solely rule the biodiesel swelling behaviour of rubber materials, but that the choice of the accelerator also plays a vital role on the swelling capacity of nitrile rubber.

Furthermore, composition *5(3/0/0.5)* also swelled less than composition *4(1/2/1.5)*, despite having a lower crosslink density. This comparison is also valid between compositions *4(1/2/1.5)* and *7(3/2/0.5)*. Compositions *5(3/0/0.5)*, *6(3/0/1.5)*, and *7(3/2/0.5)* were prepared with a high TMTD content, whereas composition *4(1/2/1.5)* was prepared with a low TMTD content. These results implied that the crosslink network formed by high amounts of the accelerator TMTD led to a more effective network that was able to restrict the swelling of nitrile rubber when it was immersed in biodiesel.

Figure 32 matches the change in mass after 700 h of immersion with the crosslink densities of the compositions. It can be clearly observed that the compositions with EV systems ended the tests less swollen than those with SEV or CV systems. Moreover, the same comments drawn for the early immersion times are also valid for samples after being immersed for 700 h. The crosslink density does not solely rule the swelling phenomenon of biodiesel: the compositions *5(3/0/0.5)* and *7(3/2/0.5)* still swelled less than composition *4(1/2/1.5)* even after 700 h, despite their lower crosslink densities.

However, not replacing the biodiesel with new non-aged biodiesel after 700 h of ageing tests may not depict actual engines conditions.

Figure 32 – Changes in mass after 700 h of immersion as a function of the crosslink densities of the nitrile rubber compositions. Numbers between brackets indicate the amount in phr of each curing system component of the formulations (TMTD/CBS/sulphur).

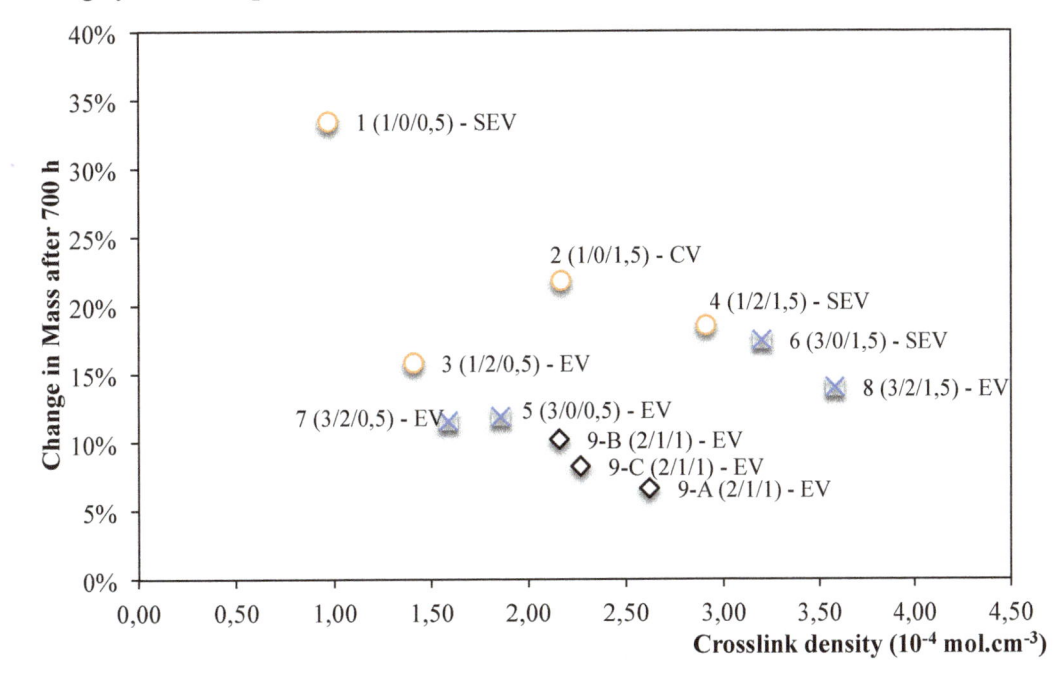

5.2.2.2 Differential scanning calorimetry (DSC)

The glass transition temperature (T_g) change-profile of the compositions before and after their immersion in soybean oil biodiesel is presented in Figure 33. It was observed that the T_g of the compositions decreased after their immersion in biodiesel. The decrease was more pronounced during the first 22 h of immersion. For longer times of immersion, the T_g stayed almost constant compared to when the samples were immersed for 22 h. The exact T_g's of the compositions before and after their immersion in soybean biodiesel are presented in Table 14.

This result corroborates the idea that the absorbed biodiesel acts as a plasticiser (TRAKARNPRUK; PORNTANGJITLIKIT, 2008), which reduces chains entanglement and eases the mobility of the polymer chains (MEYER et al., 2006). Nonetheless, the higher mobility of the polymer chains may lead to losses of mechanical properties of the materials.

Figure 33 – Glass transition temperatures (°C) of the nitrile rubber compositions before and after 22 h, 46 h, 166 h, and 700 h of immersion in soybean biodiesel at 100°C. Numbers between brackets indicate the amount in phr of each curing system component of the formulations (TMTD/CBS/sulphur).

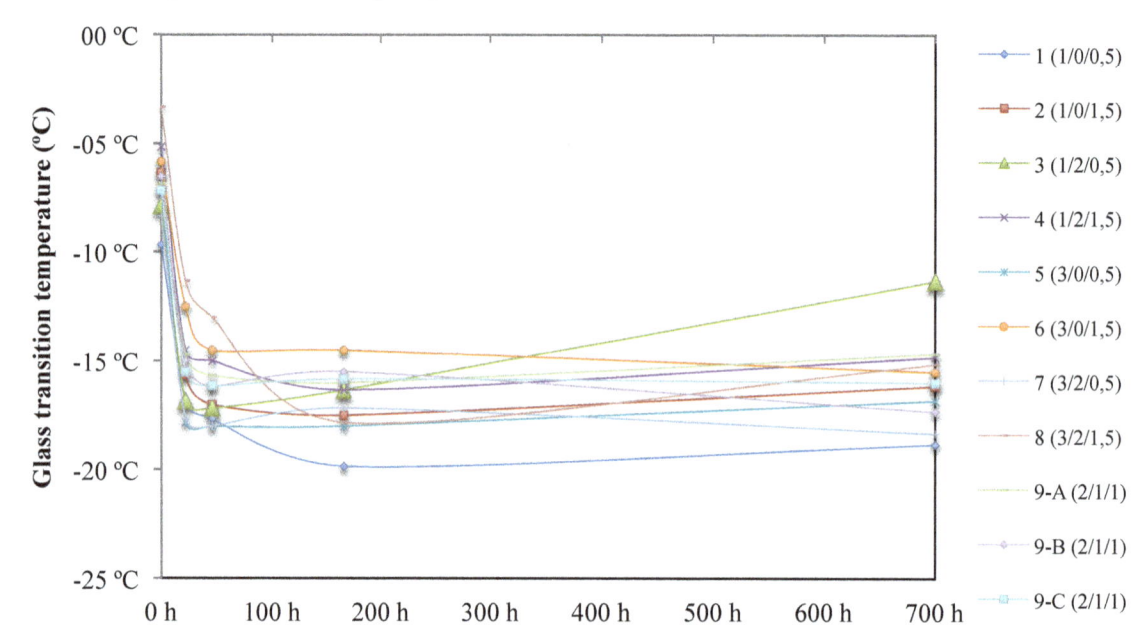

Table 14 – Glass transition temperatures (°C) of the nitrile rubber compositions before and after 22 h, 46 h, 166 h, and 700 h of immersion in soybean biodiesel at 100°C. Numbers between brackets indicate the amount in phr of each curing system component of the formulations (TMTD/CBS/sulphur).

Composition	Glass transition temperature (°C)				
	Non-immersed	After 22 h of immersion	After 46 h of immersion	After 166 h of immersion	After 700 h of immersion
1 (1/0/0.5)	-9,7	-16,8	-17,7	-19,8	-18,8
2 (1/0/1.5)	-6,3	-15,7	-17,0	-17,5	-16,2
3 (1/2/0.5)	-7,8	-16,8	-17,2	-16,3	-11,3
4 (1/2/1.5)	-5,2	-14,5	-15,0	-16,3	-14,8
5 (3/0/0.5)	-7,2	-17,8	-18,0	-18,0	-16,8
6 (3/0/1.5)	-5,8	-12,5	-14,5	-14,5	-15,5
7 (3/2/0.5)	-8,3	-17,7	-18,0	-17,2	-18,3
8 (3/2/1.5)	-3,3	-11,3	-13,0	-17,8	-15,2
9-A (2/1/1)	-6,8	-14,7	-15,7	-16,0	-14,7
9-B (2/1/1)	-6,5	-15,0	-16,2	-15,5	-17,3
9-C (2/1/1)	-7,2	-15,5	-16,2	-15,8	-16,0

5.2.2.3 Strain-stress

First, to assess the effect of the temperature n the mechanical properties of the compositions, the crosslink densities of the compositions were calculated after the heating ageing test (Figure 34). After ageing in air for 22 h, the crosslink densities of the compositions slightly increased, indicating an over-curing process; however, no degradation process was detected.

Figure 34 – Crosslink densities of the nitrile rubber compositions before and after heating ageing process for 22h at 100°C. Numbers between brackets indicate the amount in phr of each curing system component of the formulations (TMTD/CBS/sulphur).

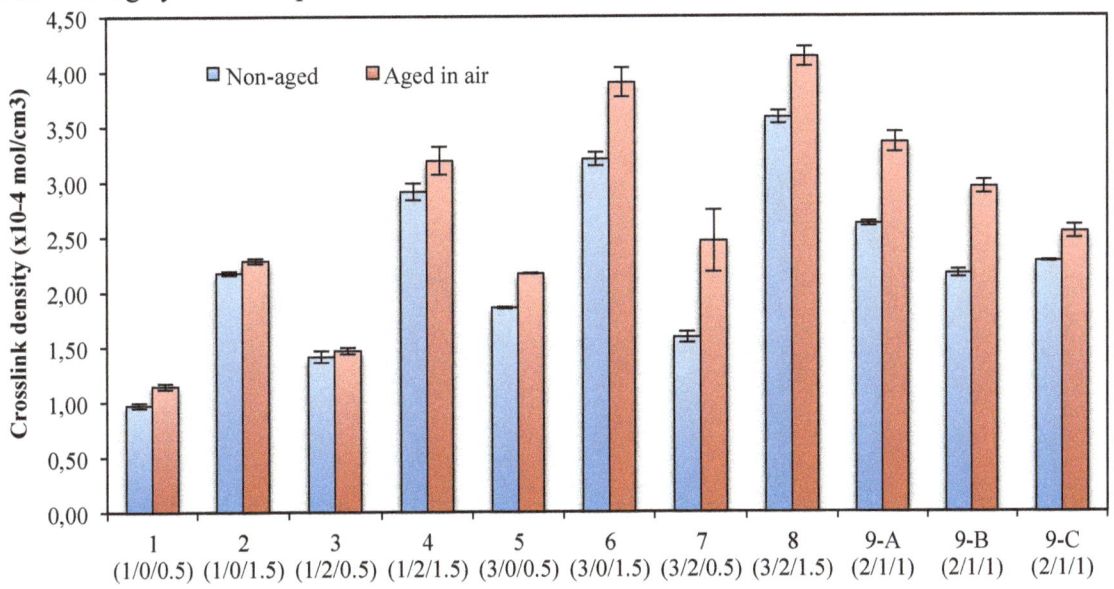

The tensile strengths of the non-aged compositions, heat-aged compositions, and compositions after 22 h of immersion in biodiesel, as well as the relative changes in tensile strength after immersion in soybean biodiesel are shown in Figure 35.

It could be noted that heating did not have much effect on the tensile strengths of the compositions, as only small changes were observed compared to the non-aged samples. Some of the compositions (*5(3/0/0.5)* and *7(3/2/0.5)*) exhibited slight increases in their tensile strengths, which was likely due to over-curing of the samples. Thus, the changes observed after immersing the samples in biodiesel could be directly related to the degradation process by biodiesel itself.

One could observe that the compositions prepared with high TMTD content even after immersion presented, on average, higher values of tensile strength. The formulation *6(3/0/1.5)* showed the lowest loss of its initial tensile strength (-10%) after immersion, and also one of the highest tensile values after immersion in biodiesel (17 MPa).

Moreover, the composition *3(1/2/0.5)*, which had the highest tensile strength before immersion, lost 61% of its initial properties and resulted in the lowest tensile strength after immersion (9 MPa).

Figure 35 – **(a)** Tensile strengths (MPa) of the nitrile rubber compositions: non-aged, after 22 h of ageing in air at 100°C, and after 22 h of immersion in soybean biodiesel at 100°C. **(b)** Relative change in tensile strengths of the nitrile rubber compositions after immersion in soybean biodiesel for 22 h at 100°C. Numbers between brackets indicate the amount in phr of each curing system component of the formulations (TMTD/CBS/sulphur).

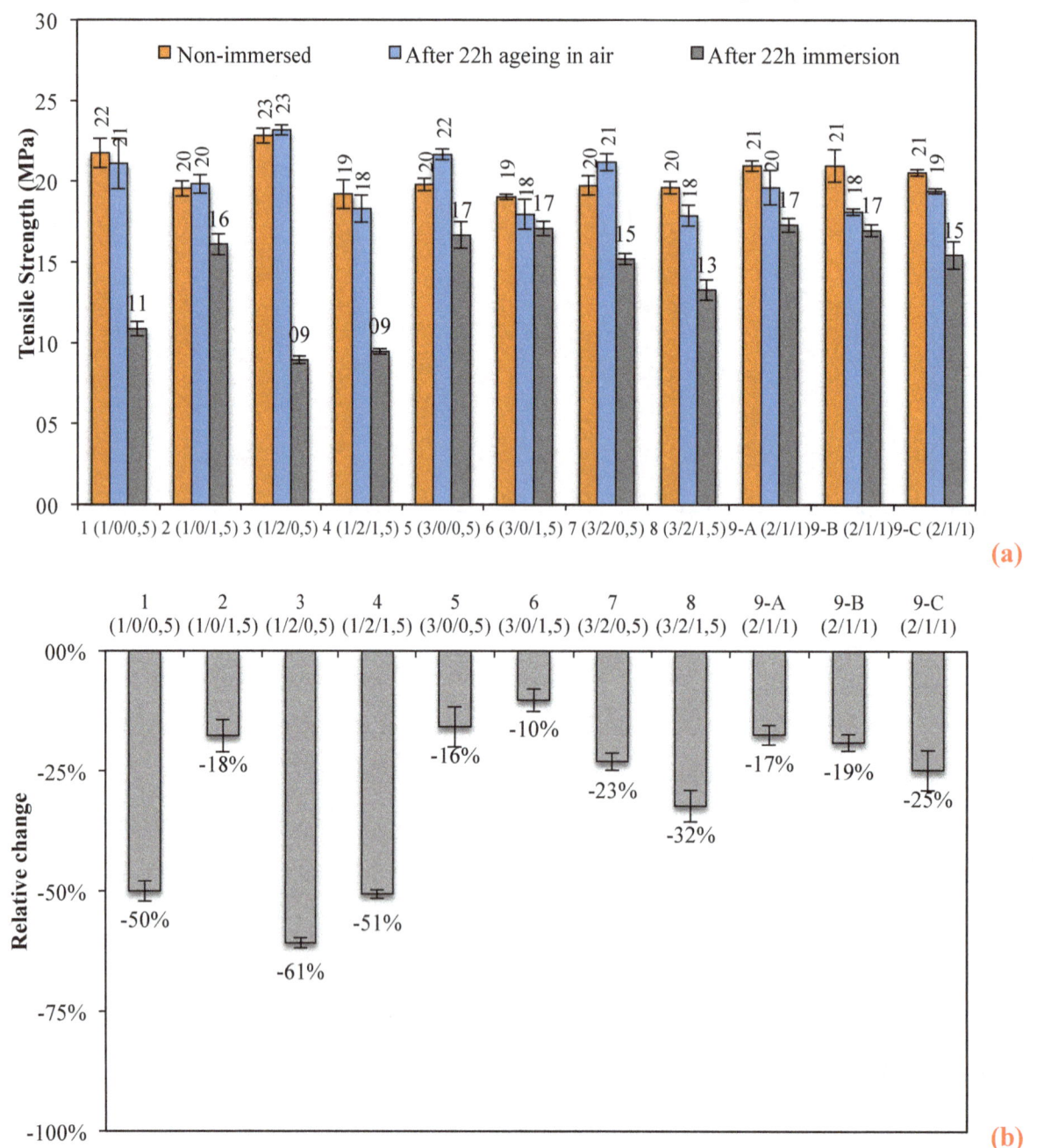

The elongation at break of the non-aged compositions, heat-aged compositions, and the compositions after 22 h of immersion in biodiesel, as well as the relative changes in elongation at break after immersion in soybean biodiesel are shown in Figure 36. The elongation at break was expected to decrease after thermal ageing, given the over-curing process that the composition underwent. The increase in the crosslink density would obviously lead to a decrease in elongation, as already discussed in section 5.2.1.2 (Figure 23).

Several authors also reported a decrease in mechanical properties after immersion in biodiesel (TRAKARNPRUK; PORNTANGJITLIKIT, 2008; HASEEB et al., 2010; HASEEB et al., 2011b; ALVES; MELLO; MEDEIROS, 2013; LINHARES et al., 2013; CORONADO et al., 2014; AKHLAGHI et al., 2015b; CH'NG et al., 2015; ZHU et al., 2015), but to different extents. These differences were related to the different testing conditions, e.g., origin of biodiesel, immersion period, immersion temperature, and grade of nitrile rubber. Most of these authors failed to provide precise information on the type of nitrile rubber that was used, which could lead to significantly different results.

Corroborating the results of the tensile strengths of the materials after immersion in biodiesel, the elongation at break values of the compositions prepared with a high TMTD content were less affected by the biofuel than those prepared with a low TMTD content.

The decreases in elongation at break and tensile strength could be credited to plasticisation of the rubber, as observed by the reduction in the Tg (in section 5.2.2.2). Biodiesel would reduce polymer chain entanglement (TRAKARNPRUK; PORNTANGJITLIKIT, 2008; CORONADO et al., 2014), which would consequently reduce the mechanical properties of the rubber compositions.

However, Akhlaghi et al. (2015b) and Haseeb et al. (2010; 2011b) suggested that the products generated from biodiesel oxidation had chemical interactions with the rubber materials, rather than simply undergoing physical interactions. Reactions could have occurred with the free double bonds present in the elastomers (C=C), causing degradation (oxidation) of the rubber and a reduction of polymer-filler interactions.

Interestingly, after the samples were immersed in biodiesel the elongation at break of the compositions prepared with a high TMTD content were higher than those after thermal ageing. One could infer that the major effect of biodiesel for these compositions was only physical, i.e., plasticisation, with little or no chemical interactions with the rubber. Moreover, the other compositions, whose elongation were highly deteriorated and resulted in a smaller value than after thermal ageing, were more affected likely due to chemical reactions, in agreement with observations reported by Akhlaghi et al. (2015b), and Haseeb et al. (2010; 2011b).

Figure 36 – **(a)** Elongation at break (%) of the nitrile rubber compositions: non-aged, after 22 h of ageing in air at 100°C, and after 22 h of immersion in soybean biodiesel at 100°C. **(b)** Relative change in the elongation at break of the nitrile rubber compositions after immersion in soybean biodiesel for 22 h at 100°C. Numbers between brackets indicate the amount in phr of each curing system component of the formulations (TMTD/CBS/sulphur).

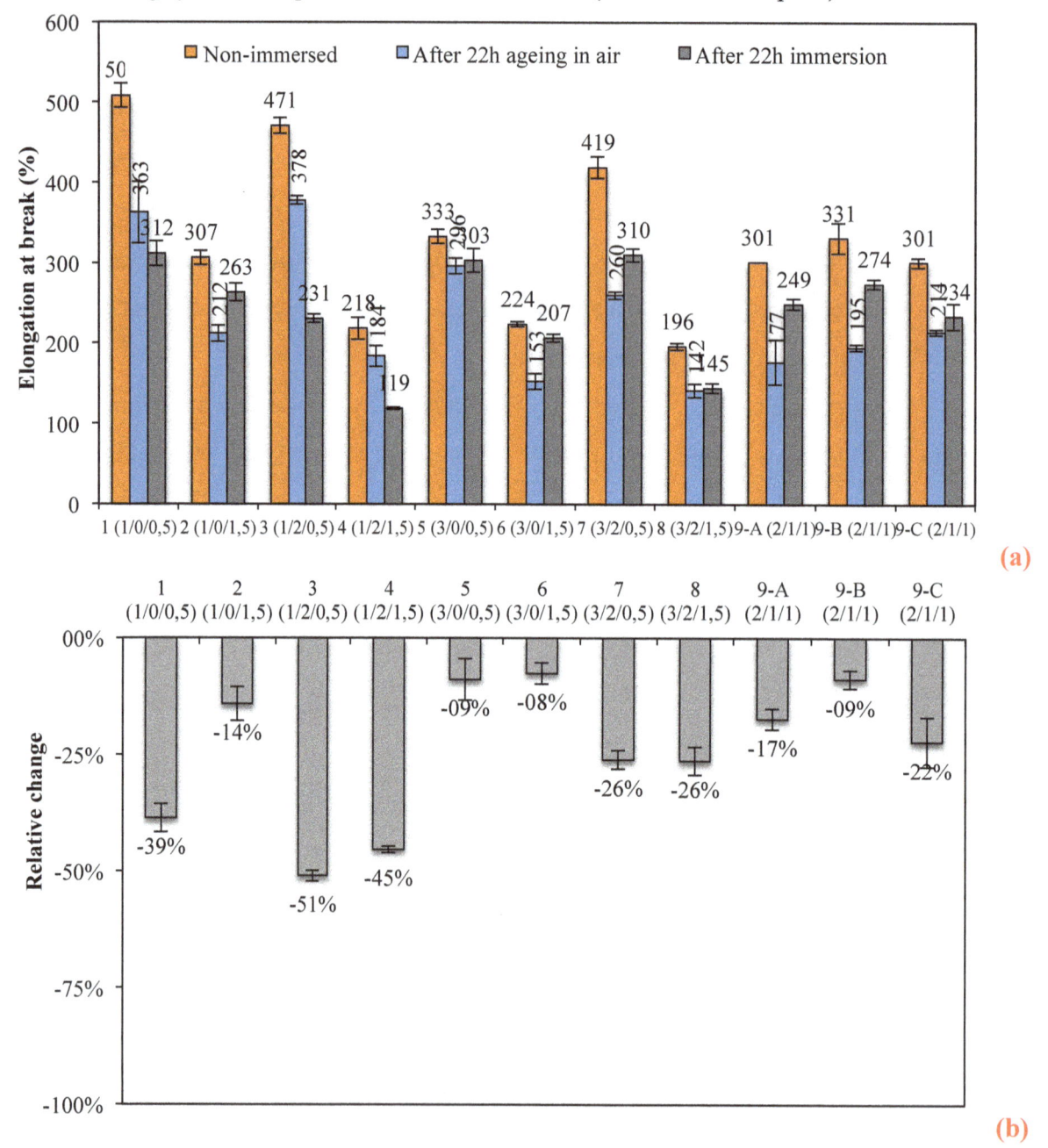

These results implied that the vulcanisation system alone (CV, SEV or EV) was not the major factor governing the mechanical resistance of nitrile rubber to biodiesel because no clear relation could be drawn. Nevertheless, the type of accelerator used seemed to play a more important role on the resistance of the samples. Compositions *4(1/2/1.5)* and *6(3/0/1,5)* had the same accelerator/sulphur ratio and consequently were considered to have SEV system. However, composition *4(1/2/1.5)* was much more degraded than composition *6(3/0/1.5)* with respect to tensile strength and elongation at break, as depicted in Figure 35 and Figure 36.

Matching the losses in tensile strength after immersion with the crosslink density (Figure 37) resulted in no obvious tendencies. However, some behavioural traits could be highlighted. Compositions *4(1/2/1.5)* and *6(3/0/1,5)* had similar crosslink density (Figure 20 in section 5.2.1.1); however the composition *4(1/2/1.5)* lost 50% of its initial properties, whereas *6(3/0/1,5)* lost only 10% of its initial properties. In addition, composition *4(1/2/1.5)* had a higher crosslink density (2.91×10^{-4} mol/cm^{-3}) than composition *5(3/0/0.5)*, whose crosslink density was 1.86×10^{-4} mol.cm^{-3}. However, the latter composition also exhibited better resistance to biodiesel, i.e., a lower change in tensile strength.

Figure 37 – Relative change in the tensile strength after immersion in soybean biodiesel for 22 h at 100°C as a function of the crosslink densities of the nitrile rubber compositions. Numbers between brackets indicate the amount in phr of each curing system component of the formulations (TMTD/CBS/sulphur).

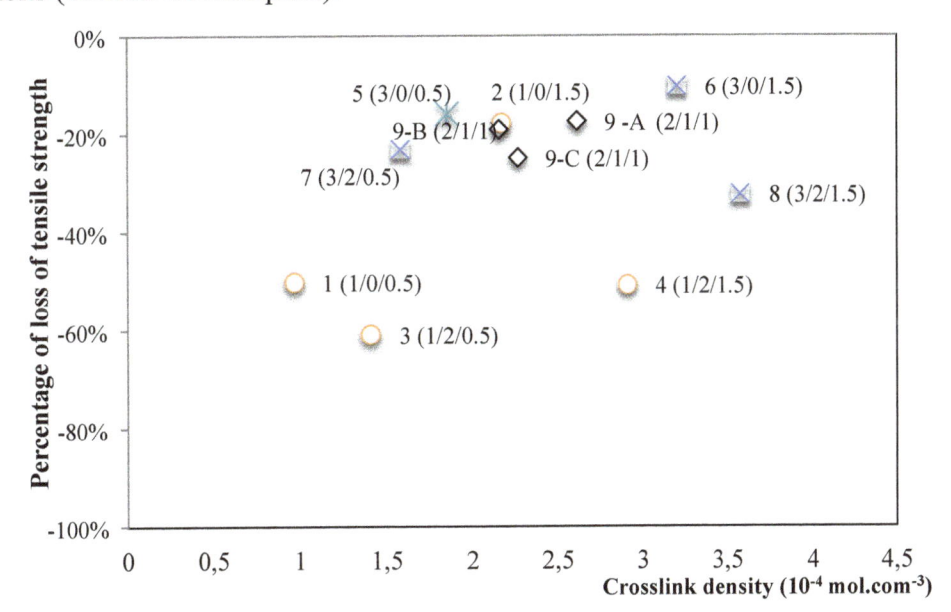

These results corroborate the suggestion proposed in section 5.2.2.1 that there is no direct relation between a material's crosslink density and the resistance to biodiesel, contrary to the propositions reported by Haseeb et al. (2010; 2011b).

To further support this suggestion, Figure 38 matches the relative loss of tensile strength after immersion in biodiesel and the accelerator/sulphur ratio that had the same ratios, namely *1(1/0/0.5)*, *4(1/2/1.5)*, and *6(3/0/1.5)*, whose ratio was 2 and had SEV systems, *3(1/2/0.5)*, and *5(3/0/0.5)*, whose ratio was 6 and had EV systems.

One can observe that the vulcanisation systems in fact did not have a role in the resistance of nitrile rubber to biodiesel. Both the compositions *1(1/0/0.5)* and *4(1/2/1.5)* were prepared with SEV systems using the same amounts of TMTD, but with different amounts of CBS and sulphur; however, they lost almost the same percentage of their initial tensile strength values. These results were compared to composition *6(3/0/1.5)*, which was also prepared with SEV system, but had high TMTD content. One can strongly suggest that the amount of TMTD had a major influence on NBR resistance to biodiesel, given that this composition lost only 10% of its initial value, despite its use of the same vulcanisation system.

Figure 38 - Relative changes in the tensile strengths after immersion in soybean biodiesel for 22 h at 100°C as a function of the vulcanisation systems of the nitrile rubber compositions. Numbers between brackets indicate the amount in phr of each curing system component of the formulations (TMTD/CBS/sulphur).

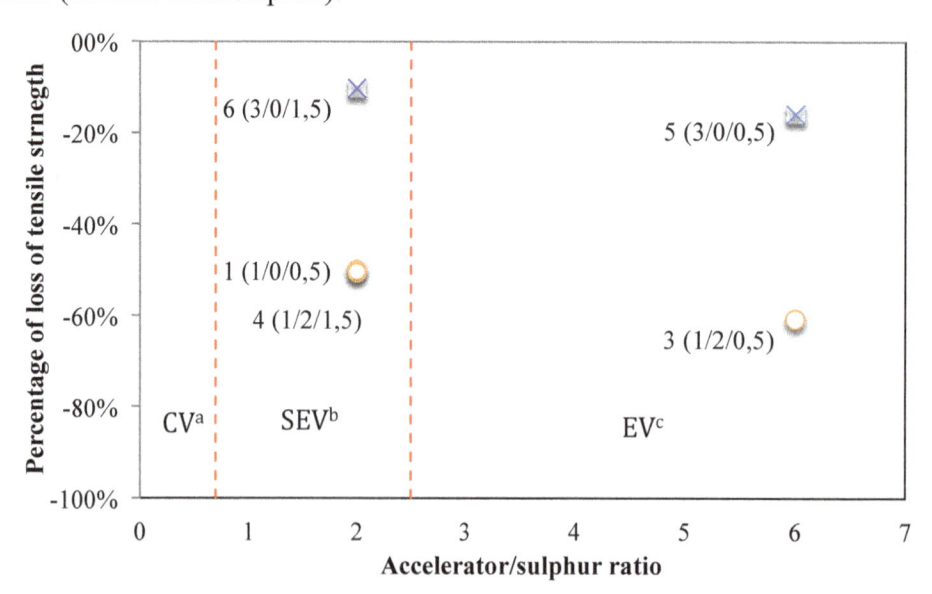

Footnote: (a)-Conventional vulcanisation system
(b)-Semi-efficient vulcanisation system
(c)-Efficient vulcanisation system

The same remarks could be made with respect to the EV system compositions. Compositions *3(1/2/0.5)* and *5(3/0/0.5)* differed from each other only in the accelerators amounts. The former was prepared with a low TMTD content and 2 phr of CBS, whereas the latter was prepared with a high TMTD content and no CBS. Compositions *5(3/0/0.5)* showed much better resistance than the composition *3(1/2/0.5)*.

5.2.2.4 Hardness

The hardness of the non-aged compositions, the compositions after 22 h of immersion in the biodiesel, and the relative change of hardness after immersion in soybean biodiesel are depicted in Figure 39.

Hardness was less affected after the immersion process than the other mechanical properties. In contrast to the other mechanical properties, high sulphur-content compositions showed an overall better performance with respect to hardness after being immersed in biodiesel. Composition *6(3/0/1.5)* had one of the smallest changes in hardness after immersion. In addition, the absolute hardness values decreased by only 7 Shore A units after immersion, which can be considered almost negligible. In agreement with these observations, ANOVA results showed that sulphur statistically influenced the changes in hardness after 22 h of immersion ($p=0.001545$), and it could be concluded from the response surface that an increase in sulphur content reduced the changes in hardness of the material after immersion.

Figure 39 – **(a)** Hardness (Shore A) of the nitrile rubber compositions: non-aged, (orange) and after 22 h of immersion in soybean biodiesel at 100°C (grey). **(b)** Relative changes in the hardness of the nitrile rubber compositions after immersion in soybean biodiesel for 22 h at 100°C. Numbers between brackets indicate the amount in phr of each curing system component of the formulations (TMTD/CBS/sulphur).

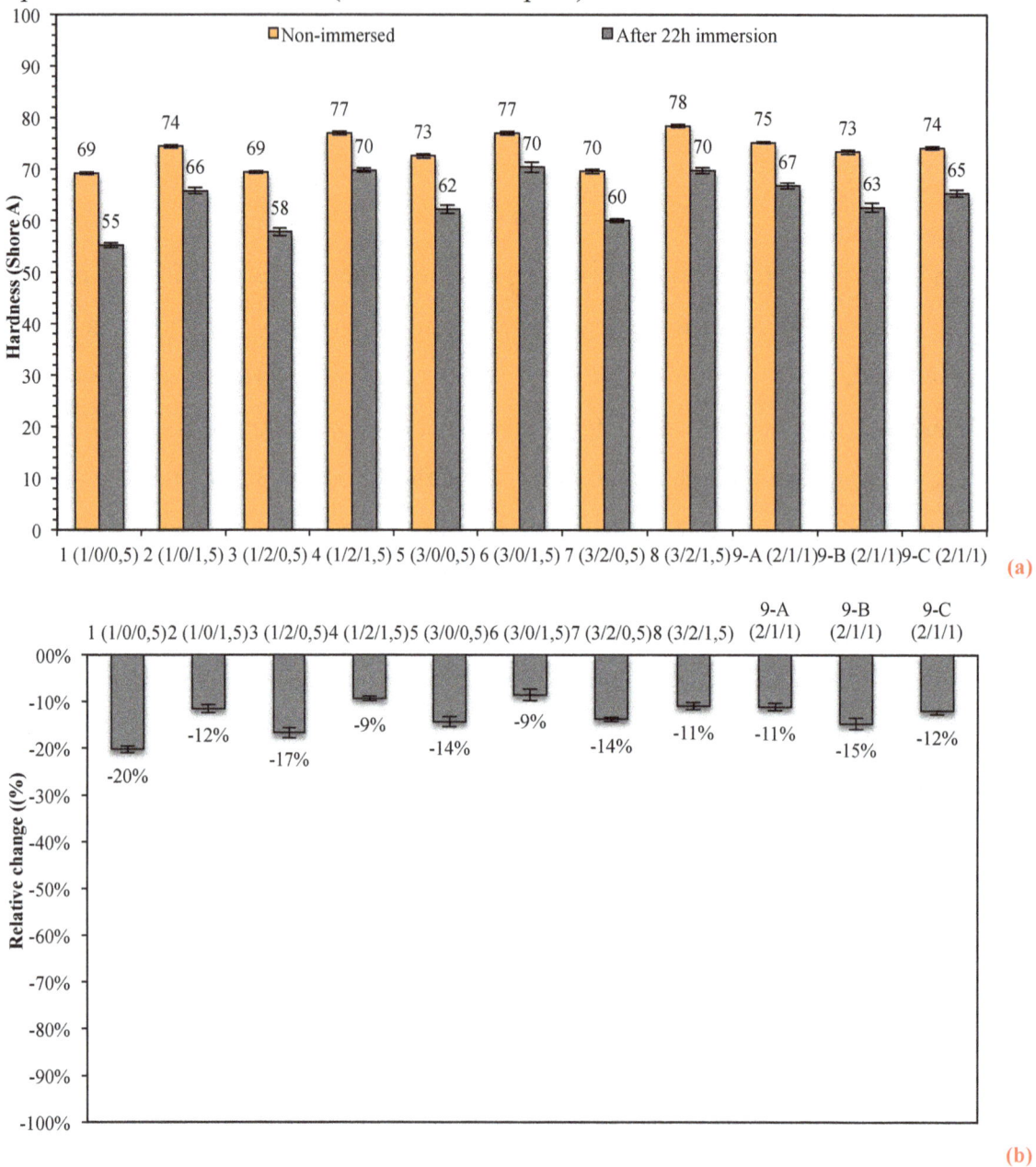

5.2.2.5 Confocal Laser Scanning Microscopy (CLSM)

At low magnifications, the CLSM photomicrographs showed that no compositions presented major changes or cracks on the fracture surfaces; these observations were expected for high-acrylonitrile-content nitrile rubbers (LINHARES et al., 2013). The CLSM photomicrographs before and after immersion of composition 3(1/2/0.5) are depicted in Figure 40.

Figure 40 – CLSM photomicrographs of the fracture surface of composition *3(1/2/0.5)*: (a) non-immersed and (b) after 22 h of immersion in soybean biodiesel at 100°C. Numbers between brackets indicate the amount in phr of each curing system component of the formulations (TMTD/CBS/sulphur).

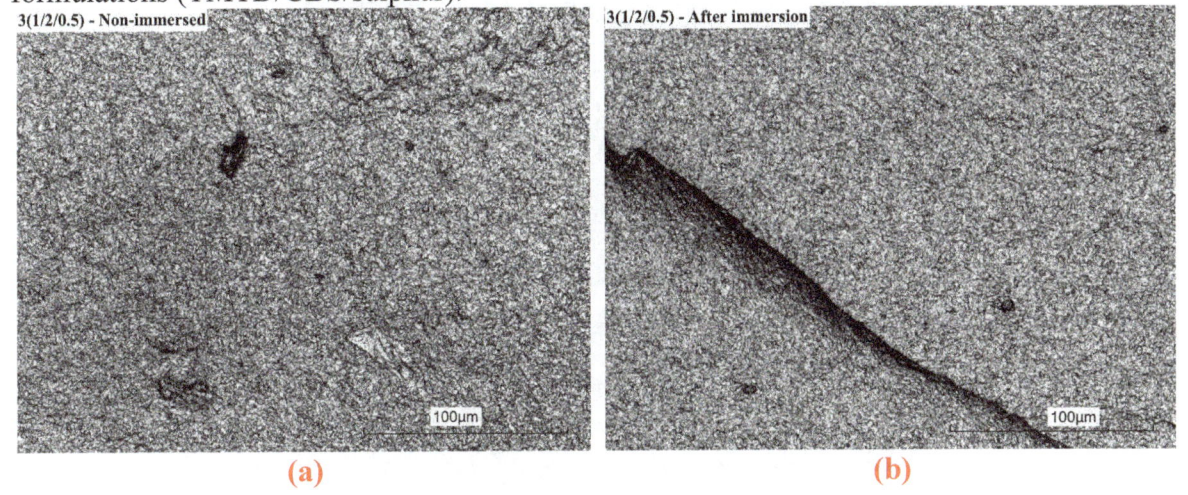

(a)   (b)

Source: LINHARES et al., 2017 (adapted)

5.2.2.6 Scanning electron microscopy (SEM)

Given the similarities of the observations, two compositions were chosen as representatives for morphology analyses using scanning electron microscopy (SEM). The SEM photomicrographs of composition *3(1/2/0.5)*, which was prepared with an EV system and low TMTD content, are depicted in Figure 41. Small voids were observed on the fracture surface of the sample after the immersion in biodiesel, which was in agreement with the remarks from Haseeb et al. (2010) and Akhlaghi et al. (2015b). Cavitation was caused by the detrimental interaction between the elastomer and biodiesel and may have induced losses in mechanical properties (section 4.2.6).

Figure 41 – SEM photomicrographs of the fracture surface of composition *3(1/2/0.5)*: (a) non-immersed and (b) after 22 h of immersion in soybean biodiesel at 100ºC. Numbers between brackets indicate the amount in phr of each curing system component of the formulations (TMTD/CBS/sulphur).

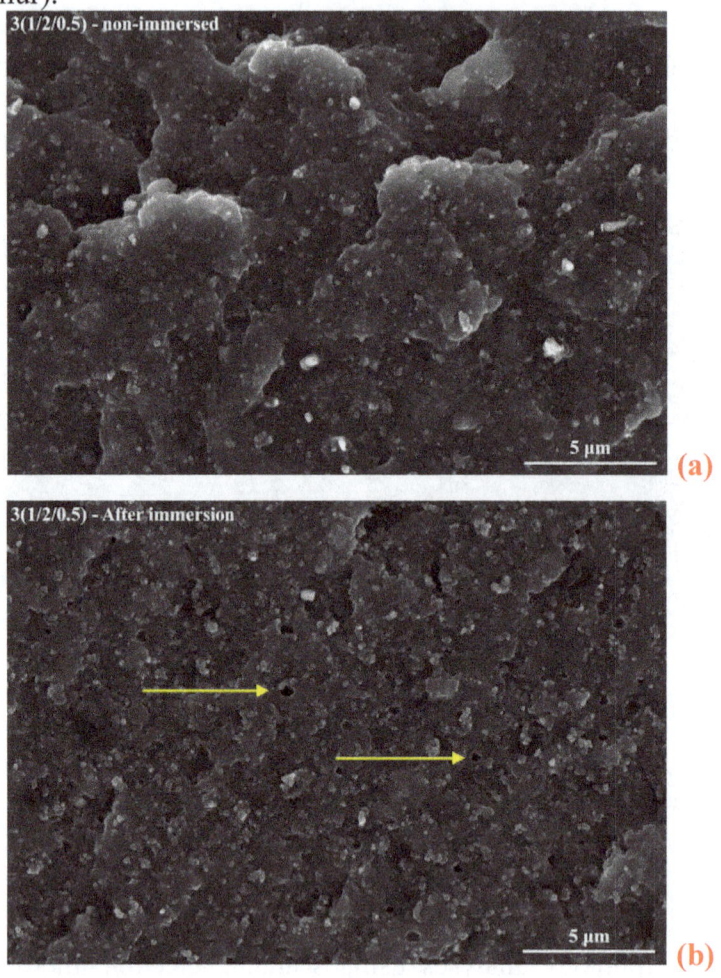

Source: LINHARES et al., 2017 (adapted)

Moreover, the SEM photomicrographs of composition *5(3/0/0.5)*, which was prepared with an EV system and had a high TMTD content, are depicted in Figure 42. The presence of crystals on the fracture surface of the non-immersed compositions has already been exhaustively discussed in sections 5.2.1.4, 5.2.1.6, and 5.2.1.7.

During the immersion, biodiesel appeared to have extracted the exceess TMTD from the compositions. However, in contrast to the composition with low TMTD content, no voids or cracks were observed on the fractured surface.

This suggests that the interaction between biodiesel and this composition was weaker and would affect the properties to a smaller extent, which corroborated the previous discussions. Additionally, one can infer that the amount of the accelerator TMTD strongly affects nitrile rubber's resistance to biodiesel.

Figure 42 - SEM photomicrographs of the fracture surface of composition *5(3/0/0.5)*: (a) non-immersed, and (b) after 22 h of immersion in soybean biodiesel at 100ºC. Numbers between brackets indicate the amount in phr of each curing system component of the formulations (TMTD/CBS/sulphur).

CONCLUSIONS

Based on the results presented herein, composition *6(3/0/1.5)*, which was formulated with 3 phr of TMTD and 1.5 phr of sulphur, could be considered the best composition with respect to its resistance to soybean biodiesel under the tested conditions. Overall, the resistance of nitrile rubber to biodiesel could be improved by increasing the acrylonitrile content of the rubber matrix and by changing the vulcanisation system of the formulation to a more appropriate one.

Given the applications of nitrile rubber-based articles, nitrile rubber does not need to be replaced with more expensive materials, for example, in some automotive applications. Nonetheless, some adjustments should be considered to facilitate the material's proper resistance to biodiesel.

After a thorough assessment of the results, the following conclusions were drawn regarding the resistance of nitrile rubber to biodiesel:

- Increasing the acrylonitrile content increases nitrile rubber's resistance to biodiesel. It was observed that nitrile rubber with medium acrylonitrile content (33%) lost on average more than 60% of its initial properties, whereas nitrile rubber with high acrylonitrile content (45%) lost an average of less than 40% of these same properties.

- Different types of crosslink networks, e.g., ionic crosslinks via metallic oxides, improved the resistance of nitrile rubber materials to biodiesel regardless of their acrylonitrile content. Carboxylated nitrile rubber, which possessed different types of crosslink networks and had a low acrylonitrile content (28%) showed resistance to biodiesel similar to nitrile rubber with a high acrylonitrile content (45%).

- Nitrile rubber with medium acrylonitrile content (33%) experienced chemical degradation by biodiesel rather than simply undergoing physical interactions. Carboxylated nitrile rubber with low acrylonitrile rubber content (28%) experienced less chemical degradation, mostly because of the different crosslink networks that were formed.

- Crosslink density alone does not directly contribute to nitrile rubber's resistance to biodiesel. Compositions of nitrile rubber with a high acrylonitrile content (45%) with similar crosslink densities demonstrated to be resistant to biodiesel to different extents. Moreover, compositions with lower crosslink densities were more resistant than compositions with higher crosslink densities.

- Furthermore, crosslink density alone does not play a role with respect to the biodiesel uptake by nitrile rubber compositions with high acrylonitrile content.

- The choice of the accelerator played a key role in determining the resistance of nitrile rubber compositions to biodiesel. Formulations with accelerator systems composed only by TMTD, or used TMTD as the primary accelerator, seemed to form more resistant crosslink networks than compositions that were formulated in the presence of CBS, or used CBS as the primary accelerator. The compositions prepared with high TMTD content (3 phr) experienced less degradation than those prepared with low amounts of TMTD (1 phr).

- The vulcanisation system (i.e., conventional (CV), semi-efficient (SEV) or efficient (EV)) alone did not play a vital role on the mechanical nitrile rubber resistance to biodiesel. Compositions with the same vulcanisation systems resisted to soybean biodiesel to different extents.

- Low-TMTD-content compositions presented voids on the fracture surface after immersion, corroborating the presence of chemical degradation processes. High-TMTD-content compositions did not show any morphological changes after immersion.

- Statistically, the amounts of TMTD, sulphur, and the combination of TMTD and sulphur influenced the tensile strengths of the compositions. Hardness was affected by the amount of TMTD, sulphur and the combination of CBS and sulphur, whereas elongation at break was affected by TMTD, sulphur, and the combination of the two accelerators. The crosslink densities of the material were influenced only by the amounts of TMTD and of sulphur.

- High-TMTD-content compositions experienced exudation of TMTD or its related decomposed components. The exudation process was time and tension dependent and could be detected by SEM, FTIR, and NMR.

- The formulations prepared were above the optimum point of the crosslink densities with respect to the tensile strengths of the compositions.

It is important to highlight that antioxidants and antiozonants are commonly added to rubber formulations, which can easily improve the resistance of rubber to aggressive media. Nevertheless, the prepared formulations did not consider the use of these additives.

SUGGESTIONS FOR FUTURE WORKS

- Assess the resistance of these nitrile rubber compositions to diesel/biodiesel blends at different concentrations and for 166h according to SAE J30;
- Assess the effect of the amount of filler on the resistance of nitrile rubber to biodiesel, as well as different types of fillers;
- Compare the resistance of prepared nitrile rubber compositions to different types of biodiesel obtained from different sources;
- Assess compositions prepared with different accelerators from different groups;
- Conduct fatigue tests with immersed and non-immersed samples according to ASTM D430:12 (De Mattia) to assess the dynamic fatigue resistance of the compositions.

REFERENCES

AFFONSO, J. E. D. S.; NUNES, R. C. R. Influence of the filler and monomer quantities in the rheometrical behaviour and crosslink density of the NBR-cellulose II composites. *Polymer Bulletin*, v. 37, p. 669-675, 1995.

AGARWAL, A. K.; GUPTA, T.; SHUKLA, P. C.; DHAR, A. Particulate emissions from biodiesel fuelled CI engines. *Energy Conversion and Management*, v. 94, p. 311-330, 2015.

AGARWAL, S.; CHHIBBER, V. K.; BHATNAGAR, A. K. Tribological behavior of diesel fuels and the effect of anti-wear additives. *Fuel*, v. 106, p. 21-29, 2013.

AKHLAGHI, S.; GEDDE, U. W.; HEDENQVIST, M. S.; BRAÑA, M. T. C.; BELLANDER, M. Deterioration of automotive rubbers in liquid biofuels: A review. *Renewable and Sustainable Energy Reviews*, v. 43, p. 1238-1248, 2015a.

AKHLAGHI, S.; HEDENQVIST, M. S.; CONDE BRAÑA, M. T.; BELLANDER, M.; GEDDE, U. W. Deterioration of acrylonitrile butadiene rubber in rapeseed biodiesel. *Polymer Degradation and Stability*, v. 111, p. 211-222, 2015b.

AKHLAGHI, S.; KALAEE, M.; MAZINANI, S.; JOWDAR, E.; NOURI, A.; SHARIF, A.; SEDAGHAT, N. Effect of zinc oxide nanoparticles on isothermal cure kinetics, morphology and mechanical properties of EPDM rubber. *Thermochimica Acta*, v. 527, p. 91-98, 2012.

ALAM, M. N.; MANDAL, S. K.; DEBNATH, S. C. Bis(N-benzyl piperazino) thiuram disulfide and dibenzothiazyl disulfide as synergistic safe accelerators in the vulcanization of natural rubber. *Journal of Applied Polymer Science*, v. 126, n. 6, p. 1830-1836, 2012a.

_____. Effect of Zinc Dithiocarbamates and Thiazole-Based Accelerators on the Vulcanization of Natural Rubber. *Rubber Chemistry and Technology*, v. 85, n. 1, p. 120-131, 2012b.

ALI, O. M.; MAMAT, R.; ABDULLAH, N. R.; ABDULLAH, A. A. Analysis of blended fuel properties and engine performance with palm biodiesel–diesel blended fuel. *Renewable Energy*, v. 86, p. 59-67, 2016.

ALICE Web. The System of Analysis of Foreign Trade Information of the Bureau of Foreign Trade, of the Brazilian Ministry of Development, Industry and Foreign Trade. Disponível em: http://aliceweb.desenvolvimento.gov.br/ >. Acesso em 15 Out. 2015.

ALVES, S. M.; MELLO, V. S.; MEDEIROS, J. S. Palm and soybean biodiesel compatibility with fuel system elastomers. *Tribology International*, v. 65, p. 74-80, 2013.

ANANDHAN, M.; KAISARE, N. S.; KANNAN, K.; VARKEY, B. "Population Balance" Model for Vulcanization of Natural Rubber with Delayed-Action Accelerator and Prevulcanization Inhibitor. *Rubber Chemistry and Technology*, v. 85, n. 2, p. 219-243, 2012.

ANDERSON, L. G. Effects of using renewable fuels on vehicle emissions. *Renewable and Sustainable Energy Reviews*, v. 47, p. 162-172, 2015.

ANDRIYANA, A.; CHAI, A. B.; VERRON, E.; JOHAN, M. R. Interaction between diffusion of palm biodiesel and large strain in rubber: Effect on stress-softening during cyclic loading. *Mechanics Research Communications*, v. 43, p. 80-86, 2012.

Agência Nacional do Petróleo, Gás Natural e Biocombustíveis. Agência Nacional do Petróleo, Gás Natural e Biocombustíveis. Disponível em: http://www.anp.gov.br >. Acesso em 15 Jul. 2015.

APREM, A. S.; JOSEPH, K.; THOMAS, S. Recent developments in crosslinking of elastomers. *Rubber Chemistry and Technology*, v. 78, p. 458-488, 2005.

ARMAS, O.; GÓMEZ, A.; RAMOS, Á. Comparative study of pollutant emissions from engine starting with animal fat biodiesel and GTL fuels. *Fuel*, v. 113, p. 560-570, 2013.

AVHAD, M. R.; MARCHETTI, J. M. A review on recent advancement in catalytic materials for biodiesel production. *Renewable and Sustainable Energy Reviews*, v. 50, p. 696-718, 2015.

AVINASH, A.; SUBRAMANIAM, D.; MURUGESAN, A. Bio-diesel—A global scenario. *Renewable and Sustainable Energy Reviews*, v. 29, p. 517-527, 2014.

BARLOW, F. W. *Rubber Compounding: principles, materials, and techniques*. 1. ed. New York: Marcel Dekker, 1988.

BERGMANN, J. C.; TUPINAMBÁ, D. D.; COSTA, O. Y. A.; ALMEIDA, J. R. M.; BARRETO, C. C.; QUIRINO, B. F. Biodiesel production in Brazil and alternative biomass feedstocks. *Renewable and Sustainable Energy Reviews*, v. 21, p. 411-420, 2013.

BERGTHORSON, J. M.; THOMSON, M. J. A review of the combustion and emissions properties of advanced transportation biofuels and their impact on existing and future engines. *Renewable and Sustainable Energy Reviews*, v. 42, p. 1393-1417, 2015.

BHATTACHARJEE, S.; BHOWMICK, A. K.; AVASTH, B. N. Degradation of hydrogenated nitrile rubber. *Polymer Degradation and Stability*, v. 31, p. 71-87, 1991.

BÖHNING, M.; NIEBERGALL, U.; ADAM, A.; STARK, W. Impact of biodiesel sorption on mechanical properties of polyethylene. *Polymer Testing*, v. 34, p. 17-24, 2014a.

_____. Influence of biodiesel sorption on temperature-dependent impact properties of polyethylene. *Polymer Testing*, v. 40, p. 133-142, 2014b.

BORDBAR, M.; BIJANZADEH, H. R.; ALIZADEH, N. 1H NMR Study of Hindered Internal Rotation of Tetramethylthiuram Disulfide in Binary Dimethyl Sulfoxide-Chloroform Mixtures. *Journal of the Chinese Chemical Society*, v. 51, p. 471-476, 2004.

BROWNSTEIN, A. M. *Renewable Motor Fuels: The Past, the Present and The Uncertain Future*. Oxford: Elsevier, 2015.

BUDRUGEAC, P. Accelerated thermal ageing of nitrile- butadiene rubber under air pressured. *Polymer Degradation and Stability*, v. 47, p. 129-132, 1995.

CELINA, M.; WISE, J.; OTTESEQ, D. K.; GILLEN, K. T.; CLOUGH, R. L. Oxidation profiles of thermally aged nitrile rubber. *Polymer Degradation and Stability*, v. 60, p. 493-504, 1998.

CH'NG, S.; ANDRIYANA, A.; TEE, Y.; VERRON, E. Effects of Carbon Black and the Presence of Static Mechanical Strain on the Swelling of Elastomers in Solvent. *Materials*, v. 8, n. 3, p. 884-898, 2015.

CHAI, A. B.; ANDRIYANA, A.; VERRON, E.; JOHAN, M. R. Mechanical characteristics of swollen elastomers under cyclic loading. *Materials & Design*, v. 44, p. 566-572, 2013.

CHAI, A. B.; ANDRIYANA, A.; VERRON, E.; JOHAN, M. R.; HASEEB, A. S. M. A. Development of a compression test device for investigating interaction between diffusion of biodiesel and large deformation in rubber. *Polymer Testing*, v. 30, n. 8, p. 867-875, 2011.

CHANDRASEKARAN, V. C. *Rubber as a Construction Material for Corrosion Protection: A comprehensive guide for process equipment designers*. 1 ed. Hoboken: John Wiley and Sons; Scrivener Publishing LLC, 2010.

CHEW, K. V.; HASEEB, A. S. M. A.; MASJUKI, H. H.; FAZAL, M. A.; GUPTA, M. Corrosion of magnesium and aluminum in palm biodiesel: A comparative evaluation. *Energy*, v. 57, p. 478-483, 2013.

CHOI, S.-S. Influence of Rubber Composition on Change of Crosslink Density of Rubber Vulcanizates with EV Cure System by Thermal Aging. *Journal of Applied Polymer Science*, v. 75, p. 1378-1384, 2000.

CHOI, S.-S.; KIM, E. A novel system for measurement of types and densities of sulfur crosslinks of a filled rubber vulcanizate. *Polymer Testing*, v. 42, p. 62-68, 2015.

CHOI, S.-S.; KIM, J.-C. Lifetime prediction and thermal aging behaviors of SBR and NBR composites using crosslink density changes. *Journal of Industrial and Engineering Chemistry*, v. 18, n. 3, p. 1166-1170, 2012.

CHONG, C. T.; NG, J.-H.; AHMAD, S.; RAJOO, S. Oxygenated palm biodiesel: Ignition, combustion and emissions quantification in a light-duty diesel engine. *Energy Conversion and Management*, v. 101, p. 317-325, 2015.

CHOUDHURY, A.; BHOWMICK, A. K.; SODDEMANN, M. Effect of organo-modified clay on accelerated aging resistance of hydrogenated nitrile rubber nanocomposites and their life time prediction. *Polymer Degradation and Stability*, v. 95, n. 12, p. 2555-2562, 2010.

CHOUGH, S.; CHANG, D. Kinetcs of sulphur vulcanization of NR, BR, SBR, and their blends using a rheometer and DSC. *Journal of Applied Polymer Science*, v. 61, n. 3, p. 449-454, 1996.

CHRISTENSEN, E.; MCCORMICK, R. L. Long-term storage stability of biodiesel and biodiesel blends. *Fuel Processing Technology*, v. 128, p. 339-348, 2014.

CIULLO, P. A.; HEWITT, N. *The Rubber Formulary*. 1. ed. New York: Noyes Publications/William Andrew Publishing,, 1999.

COATES, J. Interpretation of Infrared Spectra, A Practical Approach. In: MEYER, R. A. (Ed.). *Encyclopedia of analytical chemistry: Applications, Theory and Instrumentation* Tarzana, CA: Ramtech, 2006. p.2188.

CORAN, A. Y. Chemistry of the vulcanization and protection of elastomers: A review of the achievements. *Journal of Applied Polymer Science*, v. 87, n. 1, p. 24-30, 2003.

_____. Vulcanization. In: MARK, J. E.;ERMAN, B.;EIRICH, F. R. (Ed.). *The Science and Technology of Rubber*. USA: Academic Press, 2005. p.768.

CORONADO, M.; MONTERO, G.; VALDEZ, B.; STOYTCHEVA, M.; ELIEZER, A.; GARCÍA, C.; CAMPBELL, H.; PÉREZ, A. Degradation of nitrile rubber fuel hose by biodiesel use. *Energy*, v. 68, p. 364-369, 2014.

COSTA, H. M.; VISCONTE, L. L. Y.; NUNES, R. C. R.; FURTADO, C. R. G. Aspectos históricos da vulcanização. *Polímeros*, v. 13, p. 125-129, 2003.

CREMONEZ, P. A.; FEROLDI, M.; FEIDEN, A.; GUSTAVO TELEKEN, J.; JOSÉ GRIS, D.; DIETER, J.; DE ROSSI, E.; ANTONELLI, J. Current scenario and prospects of use of liquid biofuels in South America. *Renewable and Sustainable Energy Reviews*, v. 43, p. 352-362, 2015a.

CREMONEZ, P. A.; FEROLDI, M.; NADALETI, W. C.; DE ROSSI, E.; FEIDEN, A.; DE CAMARGO, M. P.; CREMONEZ, F. E.; KLAJN, F. F. Biodiesel production in Brazil: Current scenario and perspectives. *Renewable and Sustainable Energy Reviews*, v. 42, p. 415-428, 2015b.

CURSARU, D.-L.; BRĂNOIU, G.; RAMADAN, I.; MICULESCU, F. Degradation of automotive materials upon exposure to sunflower biodiesel. *Industrial Crops and Products*, v. 54, p. 149-158, 2014.

DATTA, R. N. *Rubber Curing Systems*. Flexys BV. Netherlands: PRESS, S. R., 2002. 150 p. Relatório Técnico.

DATTA, R. N.; HUNTINK, N. M.; DATTA, S.; TALMA, A. G. Rubber vulcanizates degradation and stabilization. *Rubber Chemistry and Technology*, v. 80, p. 436-480, 2007.

DATTA, R. N.; INGHAM, F. A. A. Rubber Aditives: compounding ingredients. In: WHITE, J. R.;DE, S. K. (Ed.). *Rubber Technologist's Handbook*. Shawbury: Rapra Technology, 2001. p.596.

DAUD, N. M.; SHEIKH ABDULLAH, S. R.; ABU HASAN, H.; YAAKOB, Z. Production of biodiesel and its wastewater treatment technologies: A review. *Process Safety and Environmental Protection*, v. 94, p. 487-508, 2015.

DEBNATH, S. C.; BASU, D. K. Studies on the Effect of Thiuram Disulfide on NR Vulcanization Accelerated by Thiazole-Based Accelerator Systems. *Journal of Applied Polymer Science*, v. 60, p. 845-855, 1996.

DELOR-JESTIN, F.; BARROIS-OUDIN, N.; CARDINET, C.; LACOSTE, J.; LEMAIRE, J. Thermal ageing of acrylonitrile-butadiene copolymer. *Polymer Degradation and Stability*, v. 70, p. 1-4, 2000.

DICK, J. S.; PAWLOWSKI, H. Applications for the curemeter maximum cure rate in rubber compound development process control and cure kinetic studies. *Polymer Testing*, v. 15, p. 207-224, 1996.

DIJKHUIS, K. A. J.; NOORDERMEER, J. W. M.; DIERKES, W. K. The relationship between crosslink system, network structure and material properties of carbon black reinforced EPDM. *European Polymer Journal*, v. 45, n. 11, p. 3302-3312, 2009.

DONDI, D.; BUTTAFAVA, A.; ZEFFIRO, A.; PALAMINI, C.; LOSTRITTO, A.; GIANNINI, L.; FAUCITANO, A. The mechanisms of the sulphur-only and catalytic vulcanization of polybutadiene: An EPR and DFT study. *European Polymer Journal*, v. 62, p. 222-235, 2015.

DUBOVSKÝ, M.; BOŽEK, M.; OLŠOVSKÝ, M. Degradation of aviation sealing materials in rapeseed biodiesel. *Journal of Applied Polymer Science*, v. 132, n. 28, p. n/a-n/a, 2015.

E. I. DU PONT DE NEMOURS and COMPANY. LYONS, D. F.; MORKEN, P. A. *Acid Resistant Fluoroelastomer compositions.* International 31 ago. 2011, 28 fev. 2013.

EKENER-PETERSEN, E.; HÖGLUND, J.; FINNVEDEN, G. Screening potential social impacts of fossil fuels and biofuels for vehicles. *Energy Policy*, v. 73, p. 416-426, 2014.

EL-NEMR, K. F. Effect of different curing systems on the mechanical and physico-chemical properties of acrylonitrile butadiene rubber vulcanizates. *Materials & Design*, v. 32, n. 6, p. 3361-3369, 2011.

ELHAMOULY, S. H.; MASOUD, M. A.; KANDIL, A. M. Influences of Accelerators on the Structures & Properties of Nitrile Butadiene Rubber. *Modern Applied Science*, v. 4, n. 4, p. 47-61, 2010.

FAN, R. L.; ZHANG, Y.; LI, F.; ZHANG, Y. X.; SUN, K.; FAN, Y. Z. Effect of high-temperature curing on the crosslink structures and dynamic mechanical properties of gum and N330-filled natural rubber vulcanizates. *Polymer Testing*, v. 20, p. 925-936, 2001.

FAZAL, M. A.; HASEEB, A. S. M. A.; MASJUKI, H. H. A critical review on the tribological compatibility of automotive materials in palm biodiesel. *Energy Conversion and Management*, v. 79, p. 180-186, 2014.

FAZAL, M. A.; JAKERIA, M. R.; HASEEB, A. S. M. A. Effect of copper and mild steel on the stability of palm biodiesel properties: A comparative study. *Industrial Crops and Products*, v. 58, p. 8-14, 2014.

FRAME, E.; MCCORMICK, R. L. *Elastomer Compatibility Testing of Renewable Diesel Fuels*. National Renewable Energy Laboratory. USA: INSTITUTE, M. R., 2005. 21 p. Relatório Técnico.

FRAME, E. A.; BESSEE, G. B.; MARBACH JR., H. W. *Biodiesel fuel technology for military application*. Southwest Research Institute. USA: CENTER, D. T. I., 1997. 309 p. Relatório Técnico.

FRANTA, I. *Elastomers and Rubber Compounding Materials: Manufacture, Properties and Applications*. 3. ed. New York: Elsevier, 1989.

GHOSH, P.; KATARE, S.; PATKAR, P.; CARUTHERS, J. M.; VENKATASUBRAMANIAN, V.; WALKER, K. A. Sulfur vulcanization of natural rubber for benzothiazole accelerated formulations: from reaction mechanisms to a rational kinetic model. *Rubber Chemistry and Technology*, v. 76, p. 592-693, 2003.

GIAKOUMIS, E. G. A statistical investigation of biodiesel physical and chemical properties, and their correlation with the degree of unsaturation. *Renewable Energy*, v. 50, p. 858-878, 2013.

GONZÁLEZ, L.; RODRÍGUEZ, A.; VALENTÍN, J. L.; MARCOS-FERNÁNDEZ, A.; POSADAS, P. Conventional and Efficient Crosslinking of Natural Rubber: Effect of Heterogeneities on the Physical Properties. *Kauschuk Gummi Kunststoffe*, v. 12, p. 638-643, 2005a.

GONZÁLEZ, L.; VALENTÍN, J. L.; FERNÁNDEZ-TORRES, A.; RODRÍGUEZ, A.; MARCOS-FERNÁNDEZ, A. Effect of the network topology on the tensile strength of natural rubber vulcanizate at elevated temperature. *Journal of Applied Polymer Science*, v. 98, n. 3, p. 1219-1223, 2005b.

GRADWELL, M. H. S.; GROOFF, D. Comparison of Tetraethyl- and Tetramethylthiuram Disulfide Vulcanization. I. Reactions in the Absence of Rubber. *Journal of Applied Polymer Science*, v. 80, p. 2292–2299, 2001.

GRAHAM, J. L.; STRIEBICH, R. C.; MYERS, K. J.; MINUS, D. K.; HARRISON III, W. E. Swelling of Nitrile Rubber by Selected Aromatics Blended in a Synthetic Jet Fuel. *Energy & Fuels*, v. 20, p. 759-765, 2006.

GUAN, Y.; ZHANG, L.-X.; ZHANG, L.-Q.; LU, Y.-L. Study on ablative properties and mechanisms of hydrogenated nitrile butadiene rubber (HNBR) composites containing different fillers. *Polymer Degradation and Stability*, v. 96, n. 5, p. 808-817, 2011.

HASEEB, A. S. M. A.; FAZAL, M. A.; JAHIRUL, M. I.; MASJUKI, H. H. Compatibility of automotive materials in biodiesel: A review. *Fuel*, v. 90, n. 3, p. 922-931, 2011a.

HASEEB, A. S. M. A.; JUN, T. S.; FAZAL, M. A.; MASJUKI, H. H. Degradation of physical properties of different elastomers upon exposure to palm biodiesel. *Energy*, v. 36, n. 3, p. 1814-1819, 2011b.

HASEEB, A. S. M. A.; MASJUKI, H. H.; SIANG, C. T.; FAZAL, M. A. Compatibility of elastomers in palm biodiesel. *Renewable Energy*, v. 35, n. 10, p. 2356-2361, 2010.

HERNÁNDEZ, M.; VALENTÍN, J. L.; LÓPEZ-MANCHADO, M. A.; EZQUERRA, T. A. Influence of the vulcanization system on the dynamics and structure of natural rubber: Comparative study by means of broadband dielectric spectroscopy and solid-state NMR spectroscopy. *European Polymer Journal*, v. 68, p. 90-103, 2015.

HILTZ, J. A.; MORCHAT, R. M.; KEOUGH, I. A. A DMTA study of the fuel resistance of elastomers *Thermochimica Acta*, v. 226, p. 143-154, 1993.

HOFMANN, W. *Rubber Techbology Handbook*. Oxford University Press, 1989.

HOMBACH, L. E.; WALTHER, G. Pareto-efficient legal regulation of the (bio)fuel market using a bi-objective optimization model. *European Journal of Operational Research*, v. 245, n. 1, p. 286-295, 2015.

IBARRA, L.; MARCOS-FERNÁNDEZ, A.; ALZORRIZ. Mechanistic approach to the curing of carboxylated nitrile rubber (XNBR) by zinc peroxide/zinc oxide. *Polymer*, v. 43, p. 1649-1655, 2002.

IBARRA, L.; RODRÍGUEZ, A.; MORA-BARRANTES, I. Crosslinking of unfilled carboxylated nitrile rubber with different systems: Influence on properties. *Journal of Applied Polymer Science*, v. 108, n. 4, p. 2197-2205, 2008.

IGNATZ-HOOVER, F.; TO, B. H. Vulcanization. In: RODGERS, B. (Ed.). *Rubber Compounding: Chemistry and Applications*. New York: MArcel Dekker, 2004. p.645.

JAKERIA, M. R.; FAZAL, M. A.; HASEEB, A. S. M. A. Influence of different factors on the stability of biodiesel: A review. *Renewable and Sustainable Energy Reviews*, v. 30, p. 154-163, 2014.

JIN, D.; ZHOU, X.; WU, P.; JIANG, L.; GE, H. Corrosion behavior of ASTM 1045 mild steel in palm biodiesel. *Renewable Energy*, v. 81, p. 457-463, 2015.

JOGI, B. F.; KULKARNI, M.; BRAHMANKAR, P. K.; RATNA, D. Some Studies on Mechanical Properties of Epoxy/CTBN/Clay based Polymer Nanocomposites (PNC). *Procedia Materials Science*, v. 5, p. 787-794, 2014.

KAZAMIA, E.; SMITH, A. G. Assessing the environmental sustainability of biofuels. *Trends Plant Sci*, v. 19, n. 10, p. 615-618, 2014.

KOVÁCS, A.; TÓTH, J.; ISAÁK, G.; KERESZTÉNYI, I. Aspects of storage and corrosion characteristics of biodiesel. *Fuel Processing Technology*, v. 134, p. 59-64, 2015.

LANJEKAR, R. D.; DESHMUKH, D. A review of the effect of the composition of biodiesel on NOx emission, oxidative stability and cold flow properties. *Renewable and Sustainable Energy Reviews*, v. 54, p. 1401-1411, 2016.

LANXESS Deutschland GmbH. NASREDDINE, V.; SODDEMANN, M. *HNBR compositions with very high filler levels having excellent processability and resistance to aggressive fluids.* International 10 set. 2009, 5 mai. 2015.

LAWANDY, S. N.; HALIM, S. F. Effect of vulcanizing system on the crosslink density of nitrile rubber compounds. *Journal of Applied Polymer Science*, v. 96, n. 6, p. 2440-2445, 2005.

LINHARES, F. N.; CORRÊA, H. L.; KHALIL, C. N.; AMORIM MOREIRA LEITE, M. C.; GUIMARÃES FURTADO, C. R. Study of the compatibility of nitrile rubber with Brazilian biodiesel. *Energy*, v. 49, p. 102-106, 2013.

LINHARES, F. N.; FURTADO, C. R. G. Compatibilidade entre biodiesel brasileiro de mamona e borracha utilizada na indústria automobilística. In: UNESCO (Ed.). *Edição 2008 do* Prêmio MERCOSUL de Ciência e Tecnologia. Brasília: UNESCO, MBC, RECyT/MERCOSUL, CNPq, Petrobras, 2008. p.33-49.

LINHARES, F.N.; KERSCH, M.; SOUSA, A.M.F.; LEITE, M.C.A.M.; ALTSTÄDT, V.; FURTADO, C.R.G. Influence of binary curing system on the nitrile rubber mechanical properties. *Macromolecular Symposia*, v. 367, p. 55-62, 2016.

LINHARES, F.N.; KERSCH, M.; NIEBERGALL, U.; LEITE, M.C.A.M.; ALTSTÄDT, V.; FURTADO, C.R.G. Effect of different sulphur-based crosslink networks on the nitrile rubber resistance to bioiesel. *Fuel*, v. 191, p. 130-139, 2017.

LIU, X.; ZHAO, S.; ZHANG, X.; LI, X.; BAI, Y. Preparation, structure, and properties of solution-polymerized styrene-butadiene rubber with functionalized end-groups and its silica-filled composites. *Polymer*, v. 55, n. 8, p. 1964-1976, 2014.

MAGRYTA, J.; DEBEK, C.; DEBEK, D. Mechanical Properties of Swelled Vulcanizates of Polar Diene Elastomers. *Journal of Applied Polymer Science*, v. 99, p. 6, 2006.

MANO, E. B.; MENDES, L. C. *Introdução a Poliméros*. 2nd. ed. São Paulo, SP: Blücher, 1999.

MANSILLA, M. A.; MARZOCCA, A. J.; MACCHI, C.; SOMOZA, A. Influence of vulcanization temperature on the cure kinetics and on the microstructural properties in natural rubber/styrene-butadiene rubber blends prepared by solution mixing. *European Polymer Journal*, v. 69, p. 50-61, 2015.

MARYKUTTY, C. V.; MATHEW, G.; MATHEW, E. J.; THOMAS, S. Studies on Novel Binary Accelerator System in Sulfur Vulcanization of Natural Rubber. *Journal of Applied Polymer Science*, v. 90, p. 3173-3182, 2003.

MARZOCCA, A. J.; MANSILLA, M. A. Analysis of network structure formed in styrene–butadiene rubber cured with sulfur/TBBS system. *Journal of Applied Polymer Science*, v. 103, n. 2, p. 1105-1112, 2007.

MATTARELLI, E.; RINALDINI, C.; SAVIOLI, T. Combustion Analysis of a Diesel Engine Running on Different Biodiesel Blends. *Energies*, v. 8, n. 4, p. 3047-3057, 2015.

MEYER, A. L.; SOUZA, G. P. D.; OLIVEIRA, S. M. D.; TOMCZAK, F. B.; WASILKOSKI, C.; PINTO., C. E. D. S. Avaliação das Propriedades Termo-Mecânicas de Borracha Nitrílica após Ensaio de Compatibilidade de acordo com ASTM D 3455. *Polímeros*, v. 16, n. 3, p. 230-234, 2006.

MOFIJUR, M.; ATABANI, A. E.; MASJUKI, H. H.; KALAM, M. A.; MASUM, B. M. A study on the effects of promising edible and non-edible biodiesel feedstocks on engine performance and emissions production: A comparative evaluation. *Renewable and Sustainable Energy Reviews*, v. 23, p. 391-404, 2013a.

MOFIJUR, M.; HAZRAT, M. A.; RASUL, M. G.; MAHMUDUL, H. M. Comparative Evaluation of Edible and Non-edible Oil Methyl Ester Performance in a Vehicular Engine. *Energy Procedia*, v. 75, p. 37-43, 2015.

MOFIJUR, M.; MASJUKI, H. H.; KALAM, M. A.; ATABANI, A. E.; SHAHABUDDIN, M.; PALASH, S. M.; HAZRAT, M. A. Effect of biodiesel from various feedstocks on combustion characteristics, engine durability and materials compatibility: A review. *Renewable and Sustainable Energy Reviews*, v. 28, p. 441-455, 2013b.

MOHR, S. H.; WANG, J.; ELLEM, G.; WARD, J.; GIURCO, D. Projection of world fossil fuels by country. *Fuel*, v. 141, p. 120-135, 2015.

MOSER, B. R. Fuel property enhancement of biodiesel fuels from common and alternative feedstocks via complementary blending. *Renewable Energy*, v. 85, p. 819-825, 2016.

MOSTAFA, A.; ABOUEL-KASEM, A.; BAYOUMI, M. R.; EL-SEBAIE, M. G. The influence of CB loading on thermal aging resistance of SBR and NBR rubber compounds under different aging temperature. *Materials & Design*, v. 30, n. 3, p. 791-795, 2009.

MOVAHED, S. O.; ANSARIFAR, A.; MIRZAIE, F. Effect of various efficient vulcanization cure systems on the compression set of a nitrile rubber filled with different fillers. *Journal of Applied Polymer Science*, v. 132, n. 8, p. n/a-n/a, 2015.

NANJING JINSANLI RUBBER & PLASTIC CO., LTD. XU, H.; RUI, Q. *Low-temperature-resistant and novel-fuel-resistant low-pressure-change fluorine rubber goss rubber and preparation method thereof.* International [s.d.], 20 jun. 2012.

NASIR, M.; G. K. TEH, G. K. The effects of various types of crosslinks on the physical properties of natural rubber. *European Polymer Journal*, v. 24, p. 733-736, 1988.

NICOLAU, A.; LUTCKMEIER, C. V.; SAMIOS, D.; GUTTERRES, M.; PIATNICK, C. M. S. The relation between lubricity and electrical properties of low sulfur diesel and diesel/biodiesel blends. *Fuel*, v. 117, p. 26-32, 2014.

NIEUWENHUIZEN, P. J. Zinc accelerator complexes. versatile homogeneous catalysts in sulfur vulcanization. *Applied Catalysis A General*, v. 207, p. 55-68, 2001.

NIEUWENHUIZEN, P. J.; DUIN, M. V.; HAASNOOT, J. G.; REEDIJK, J.; MCGILL, W. J. The Limiting Value of ZDMC Formation: New Insight into the Reaction of ZnO and TMTD. *Journal of Applied Polymer Science*, v. 73, p. 1247–1257, 1999.

NIYOGI, U. K. *Polymer Additives and Compounding: Additives for Rubbers*. Division of Material Science. Delhi, India: RESEARCH, S. R. I. F. I., 2007. 30 p. Relatório Técnico.

ODIAN, G. *Principles of Polymerization*. 4th. ed. Hoboken, NJ: John Wiley & Sons, 2004.

OECD. Organisation for Economic Co-operation and Development. Disponível em: http://stats.oecd.org/ >. Acesso em 31 Jul. 2015.

OLIVEIRA, I. T. D.; PACHECO, E. B. A. V.; VISCONTE, L. L. Y.; OLIVEIRA, M. R. L.; RUBINGER, M. M. M. Efeito de um novo acelerador de vulcanização nas propriedades reométricas de composições de borracha nitrílica com diferentes teores de AN. *Polímeros*, v. 20, p. 366-370, 2010.

ÖZENER, O.; YÜKSEK, L.; ERGENÇ, A. T.; ÖZKAN, M. Effects of soybean biodiesel on a DI diesel engine performance, emission and combustion characteristics. *Fuel*, v. 115, p. 875-883, 2014.

PAZUR, R. J.; CORMIER, J. G. The Effect of Acrylonitrile Content on the Thermo-Oxidative Aging of Nitrile Rubber. *Rubber Chemistry and Technology*, v. 87, n. 1, p. 53-69, 2014.

PRIME, R. B. Thermosets. In: TURI, E. (Ed.). *Thermal Characterization of Polymeric Materials*. New York: Academic Press, 1981. p.972.

PULLEN, J.; SAEED, K. Experimental study of the factors affecting the oxidation stability of biodiesel FAME fuels. *Fuel Processing Technology*, v. 125, p. 223-235, 2014.

RAHMAN, M. S.; AL-MARHUBI, I. M.; AL-MAHROUQI, A. Measurement of glass transition temperature by mechanical (DMTA), thermal (DSC and MDSC), water diffusion and density methods: A comparison study. *Chemical Physics Letters*, v. 440, n. 4-6, p. 372-377, 2007.

RASHED, M. M.; KALAM, M. A.; MASJUKI, H. H.; MOFIJUR, M.; RASUL, M. G.; ZULKIFLI, N. W. M. Performance and emission characteristics of a diesel engine fueled with palm, jatropha, and moringa oil methyl ester. *Industrial Crops and Products*, v. 79, p. 70-76, 2016.

RESTREPO-FLÓREZ, J.-M.; BASSI, A.; REHMANN, L.; THOMPSON, M. R. Investigation of biofilm formation on polyethylene in a diesel/biodiesel fuel storage environment. *Fuel*, v. 128, p. 240-247, 2014.

RIZWANUL FATTAH, I. M.; MASJUKI, H. H.; KALAM, M. A.; HAZRAT, M. A.; MASUM, B. M.; IMTENAN, S.; ASHRAFUL, A. M. Effect of antioxidants on oxidation stability of biodiesel derived from vegetable and animal based feedstocks. *Renewable and Sustainable Energy Reviews*, v. 30, p. 356-370, 2014.

ROCHA, T. C. J.; SOARES, B. G.; COUTINHO, F. M. B. Principais copolímeros elastoméricos à base de butadieno utilizados na indústria automobilística. *Polímeros*, v. 17, n. 4, p. 299-307, 2007.

SAHOO, S.; MAITI, M.; GANGULY, A.; JACOB GEORGE, J.; BHOWMICK, A. K. Effect of zinc oxide nanoparticles as cure activator on the properties of natural rubber and nitrile rubber. *Journal of Applied Polymer Science*, v. 105, n. 4, p. 2407-2415, 2007.

SAJID, Z.; KHAN, F.; ZHANG, Y. Process simulation and life cycle analysis of biodiesel production. *Renewable Energy*, v. 85, p. 945-952, 2016.

SANGEETA; MOKA, S.; PANDE, M.; RANI, M.; GAKHAR, R.; SHARMA, M.; RANI, J.; BHASKARWAR, A. N. Alternative fuels: An overview of current trends and scope for future. *Renewable and Sustainable Energy Reviews*, v. 32, p. 697-712, 2014.

SANTOS, E. M.; PIOVESAN, N. D.; DE BARROS, E. G.; MOREIRA, M. A. Low linolenic soybeans for biodiesel: Characteristics, performance and advantages. *Fuel*, v. 104, p. 861-864, 2013.

SAXENA, P.; JAWALE, S.; JOSHIPURA, M. H. A Review on Prediction of Properties of Biodiesel and Blends of Biodiesel. *Procedia Engineering*, v. 51, p. 395-402, 2013.

SERQUEIRA, D. S.; FERNANDES, D. M.; CUNHA, R. R.; SQUISSATO, A. L.; SANTOS, D. Q.; RICHTER, E. M.; MUNOZ, R. A. A. Influence of blending soybean, sunflower, colza, corn, cottonseed, and residual cooking oil methyl biodiesels on the oxidation stability. *Fuel*, v. 118, p. 16-20, 2014.

SERRANO, M.; BOUAID, A.; MARTÍNEZ, M.; ARACIL, J. Oxidation stability of biodiesel from different feedstocks: Influence of commercial additives and purification step. *Fuel*, v. 113, p. 50-58, 2013.

SHAHIR, V. K.; JAWAHAR, C. P.; SURESH, P. R. Comparative study of diesel and biodiesel on CI engine with emphasis to emissions—A review. *Renewable and Sustainable Energy Reviews*, v. 45, p. 686-697, 2015.

SILVERSTEIN, R. M.; WEBSTER, F. X.; KIEMLE, D. J. *Spectrometric identification of organic compounds*. 7th. ed. Hoboken, NJ: John Wiley & Sons, 2005.

SINGER, P.; RÜHE, J. On the mechanism of deposit formation during thermal oxidation of mineral diesel and diesel/biodiesel blends under accelerated conditions. *Fuel*, v. 133, p. 245-252, 2014.

SORATE, K. A.; BHALE, P. V. Impact of biodiesel on fuel system materials durability. *Journal of Scientific & Industrial Research*, v. 72, p. 48-57, 2013.

SORATE, K. A.; BHALE, P. V. Biodiesel properties and automotive system compatibility issues. *Renewable and Sustainable Energy Reviews*, v. 41, p. 777-798, 2015.

SORATE, K. A.; BHALE, P. V.; DHAOLAKIYA, B. Z. A Material Compatibility Study of Automotive Elastomers with high FFA based Biodiesel. *Energy Procedia*, v. 75, p. 105-110, 2015.

SOUSA, A. M. F.; PERES, A. C. C.; NUNES, R. C. R.; VISCONTE, L. L. Y.; FURTADO, C. R. G. Resin-vulcanized NBR: Suitability of rheometric parameters for the calculation of cure kinetic constants. *Journal of Applied Polymer Science*, v. 84, n. 3, p. 505-513, 2002.

SPRENGER, S.; KOTHMANN, M. H.; ALTSTAEDT, V. Carbon fiber-reinforced composites using an epoxy resin matrix modified with reactive liquid rubber and silica nanoparticles. *Composites Science and Technology*, v. 105, p. 86-95, 2014.

STATISTA. The statistic Portal: Statistics and Studies from more than 18000 sources. Disponível em: http://www.statista.com/statistics/271472/biodiesel-production-in-selected-countries/ >. Acesso em 31 Jul. 2015.

SUSAMMA, A. P.; MINI, V. T. E.; KURIAKOSE, A. P. Studies on Novel Binary Accelerator System in Sulfur Vulcanization of Natural Rubber. *Journal of Applied Polymer Science*, v. 79, p. 1-8, 2001.

TERRY, B. *Impact of Biodiesel on Fuel System Component Durability NREL/TP-540-39130*. The Associated Octel Company Limited. USA: LIMITED, T. A. O. C., 2005. 149 p. Relatório Técnico.

THOMPSON, M. R.; MU, B.; EWASCHUK, C. M.; CAI, Y.; OXBY, K. J.; VLACHOPOULOS, J. Long term storage of biodiesel/petrol diesel blends in polyethylene fuel tanks. *Fuel*, v. 108, p. 771-779, 2013.

TIANJIN PENGLING RUBBER HOSE CO., LTD. HONGQI, Z.; JIAN, Y. *Biodiesel-resistant acrylate rubber composition*. International [s.d.], 05 fev. 2014.

TIANJIN PENGLING RUBBER HOSE CO., LTD. ZHANG, H.; YANG, J. *Biodiesel-resistant nitrile rubber*. International [s.d.], 13 abr. 2011.

TORREGROSA, A. J.; BROATCH, A.; PLÁ, B.; MÓNICO, L. F. Impact of Fischer–Tropsch and biodiesel fuels on trade-offs between pollutant emissions and combustion noise in diesel engines. *Biomass and Bioenergy*, v. 52, p. 22-33, 2013.

TRAKARNPRUK, W.; PORNTANGJITLIKIT, S. Palm oil biodiesel synthesized with potassium loaded calcined hydrotalcite and effect of biodiesel blend on elastomer properties. *Renewable Energy*, v. 33, n. 7, p. 1558-1563, 2008.

TUDU, K.; MURUGAN, S.; PATEL, S. K. Evaluation of a DI diesel engine run on a tyre derived fuel-diesel blend. *Journal of the Energy Institute*, 2015.

ULLAH, F.; DONG, L.; BANO, A.; PENG, Q.; HUANG, J. Current advances in catalysis toward sustainable biodiesel production. *Journal of the Energy Institute*, 2015.

UZUN, A. Air mass flow estimation of diesel engines using neural network. *Fuel*, v. 117, p. 833-838, 2014.

WALKER, F. J. Effects of Bio-Fuels on Common Static Sealing Elastomers. *Rubber Chemistry and Technology*, v. 82, p. 369-378, 2009.

WOO, C. S.; PARK, H. S. Useful lifetime prediction of rubber component. *Engineering Failure Analysis*, v. 18, n. 7, p. 1645-1651, 2011.

XIONG, Y.; CHEN, G.; GUO, S.; LI, G. Lifetime prediction of NBR composite sheet in aviation kerosene by using nonlinear curve fitting of ATR-FTIR spectra. *Journal of Industrial and Engineering Chemistry*, v. 19, n. 5, p. 1611-1616, 2013.

YAAKOB, Z.; NARAYANAN, B. N.; PADIKKAPARAMBIL, S.; UNNI K, S.; AKBAR P, M. A review on the oxidation stability of biodiesel. *Renewable and Sustainable Energy Reviews*, v. 35, p. 136-153, 2014.

YANG, R.; ZHAO, J.; LIU, Y. Oxidative degradation products analysis of polymer materials by pyrolysis gas chromatography–mass spectrometry. *Polymer Degradation and Stability*, v. 98, n. 12, p. 2466-2472, 2013.

YASIN, T.; AHMED, S.; YOSHII, F.; MAKUUCHI, K. Effect of acrylonitrile content on physical properties of electron beam irradiated acrylonitrile–butadiene rubber. *Reactive and Functional Polymers*, v. 57, n. 2-3, p. 113-118, 2003.

YUNUS KHAN, T. M.; ATABANI, A. E.; BADRUDDIN, I. A.; BADARUDIN, A.; KHAYOON, M. S.; TRIWAHYONO, S. Recent scenario and technologies to utilize non-edible oils for biodiesel production. *Renewable and Sustainable Energy Reviews*, v. 37, p. 840-851, 2014.

YUNUS KHAN, T. M.; BADRUDDIN, I. A.; BADARUDIN, A.; BANAPURMATH, N. R.; SALMAN AHMED, N. J.; QUADIR, G. A.; AL-RASHED, A. A. A.; KHALEED, H. M. T.; KAMANGAR, S. Effects of engine variables and heat transfer on the performance of biodiesel fueled IC engines. *Renewable and Sustainable Energy Reviews*, v. 44, p. 682-691, 2015.

ZHANG, X.; LI, L.; WU, Z.; HU, Z.; ZHOU, Y. *Material Compatibilities of biodiesels with elastomers, metals and plastics in a diesel engine*. SAE International. USA: INTERNATIONAL, S., 2009. 10 p. Relatório Técnico.

ZHANG, Z.-H.; BALASUBRAMANIAN, R. Influence of butanol–diesel blends on particulate emissions of a non-road diesel engine. *Fuel*, v. 118, p. 130-136, 2014.

ZHAO, J.; YANG, R.; IERVOLINO, R.; BARBERA, S. Changes of Chemical Structure and Mechanical Property Levels during Thermo-Oxidative Aging of Nbr. *Rubber Chemistry and Technology*, v. 86, n. 4, p. 591-603, 2013.

ZHU, L.; CHEUNG, C. S.; ZHANG, W. G.; HUANG, Z. Compatibility of different biodiesel composition with acrylonitrile butadiene rubber (NBR). *Fuel*, v. 158, p. 288-292, 2015.

APPENDIX A

Table 15 – Main accelerators used in sulphur vulcanisation of elastomer, their classifications and their chemical structures.

Generic chemical designation	Chemical name	Chemical structure
Thiazole	2-mercaptobenzothiazole	
	2-2'-dithiobenzothiazole	
Sulphenamide	N-t-butyl benzothiazole-2-sulphenamide	
	N-cyclohexyl benzothiazole-2-sulphenamide	
Thiuram	Tetramethyl thiuram monosulphide	
	Tetramethyl thiuram disulphide	
Dithiocarbamate	Zinc dimethyldithiocarbamate	
	Zinc diethyldithiocarbamate	

Source: DATTA; INGHAM, 2001; COSTA et al., 2003; GHOSH et al., 2003; APREM; JOSEPH; THOMAS, 2005; CORAN, 2005; ALAM; MANDAL; DEBNATH, 2012.

APPENDIX B

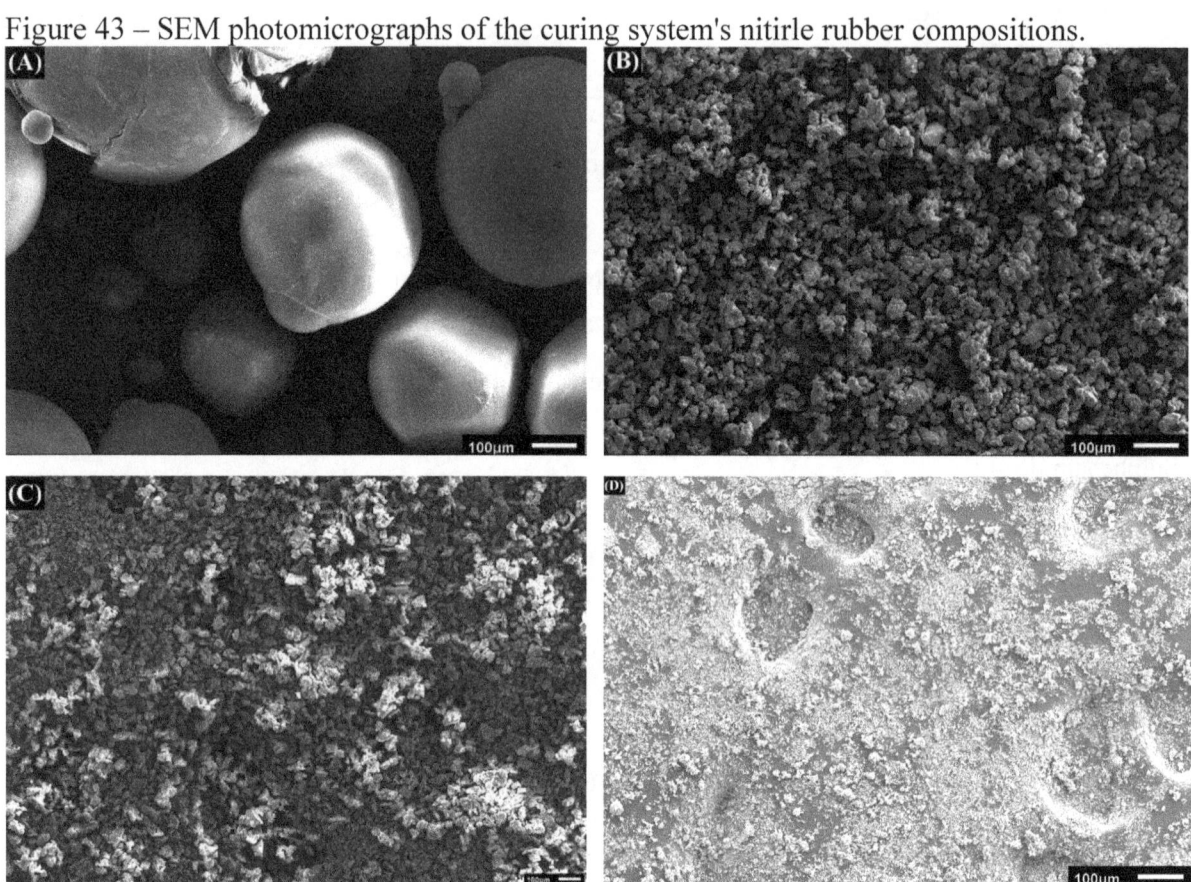

Figure 43 – SEM photomicrographs of the curing system's nitirle rubber compositions.

Footnote: (a) – stearic acid; (b) – sulphur; (c) – TMTD; (d) – zinc oxide.

APPENDIX C

Figure 44 – FTIR spectrum of pure TMTD.

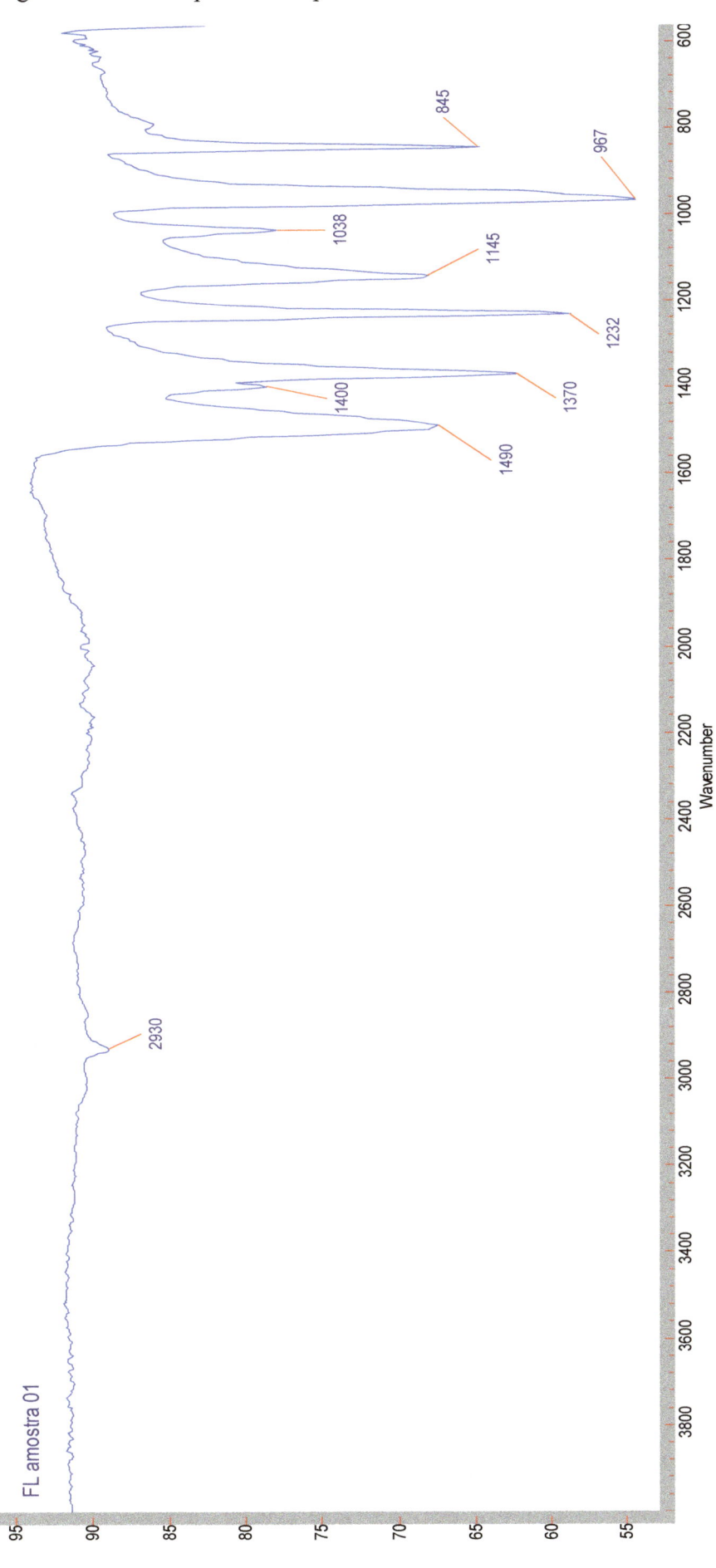

Figure 45 – FTIR spectrum of pure CBS.

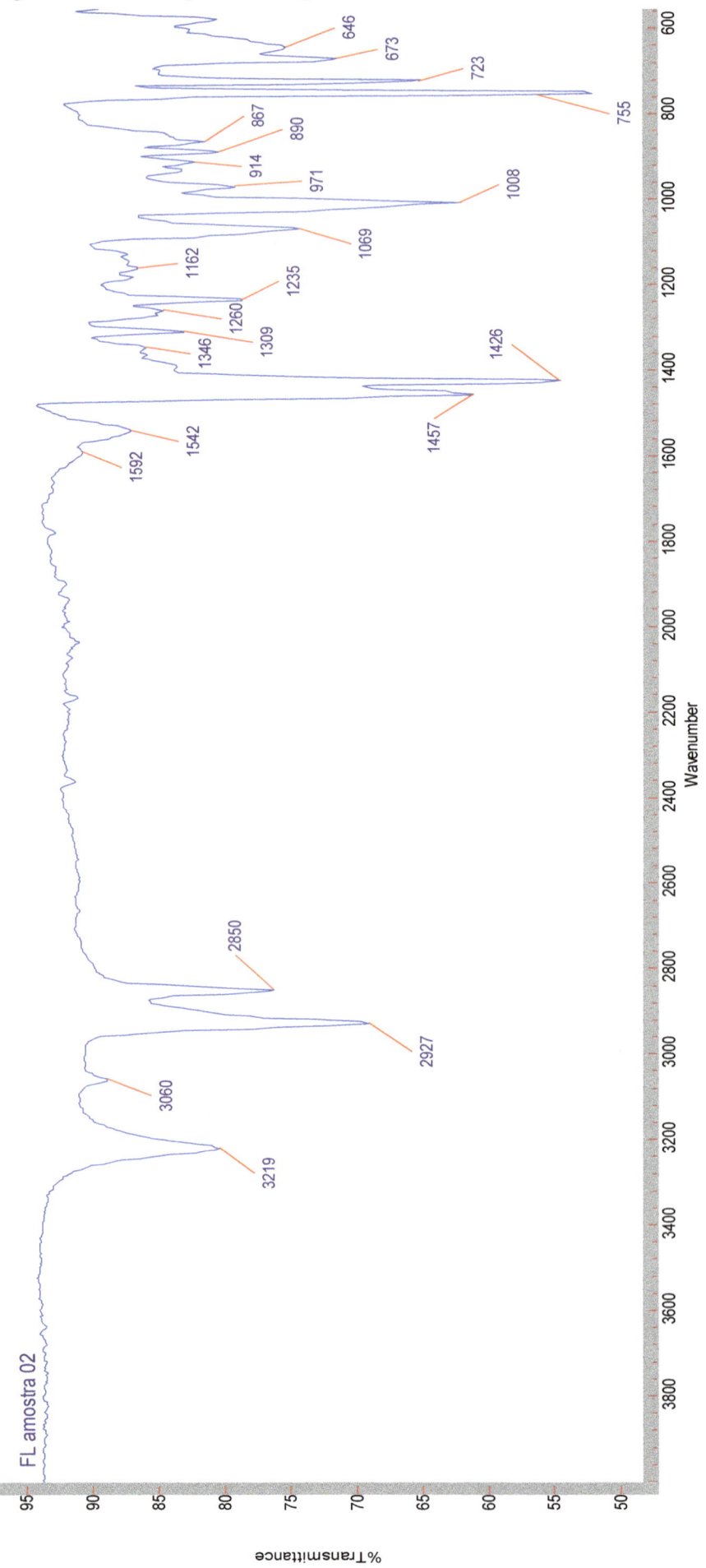

APPENDIX D

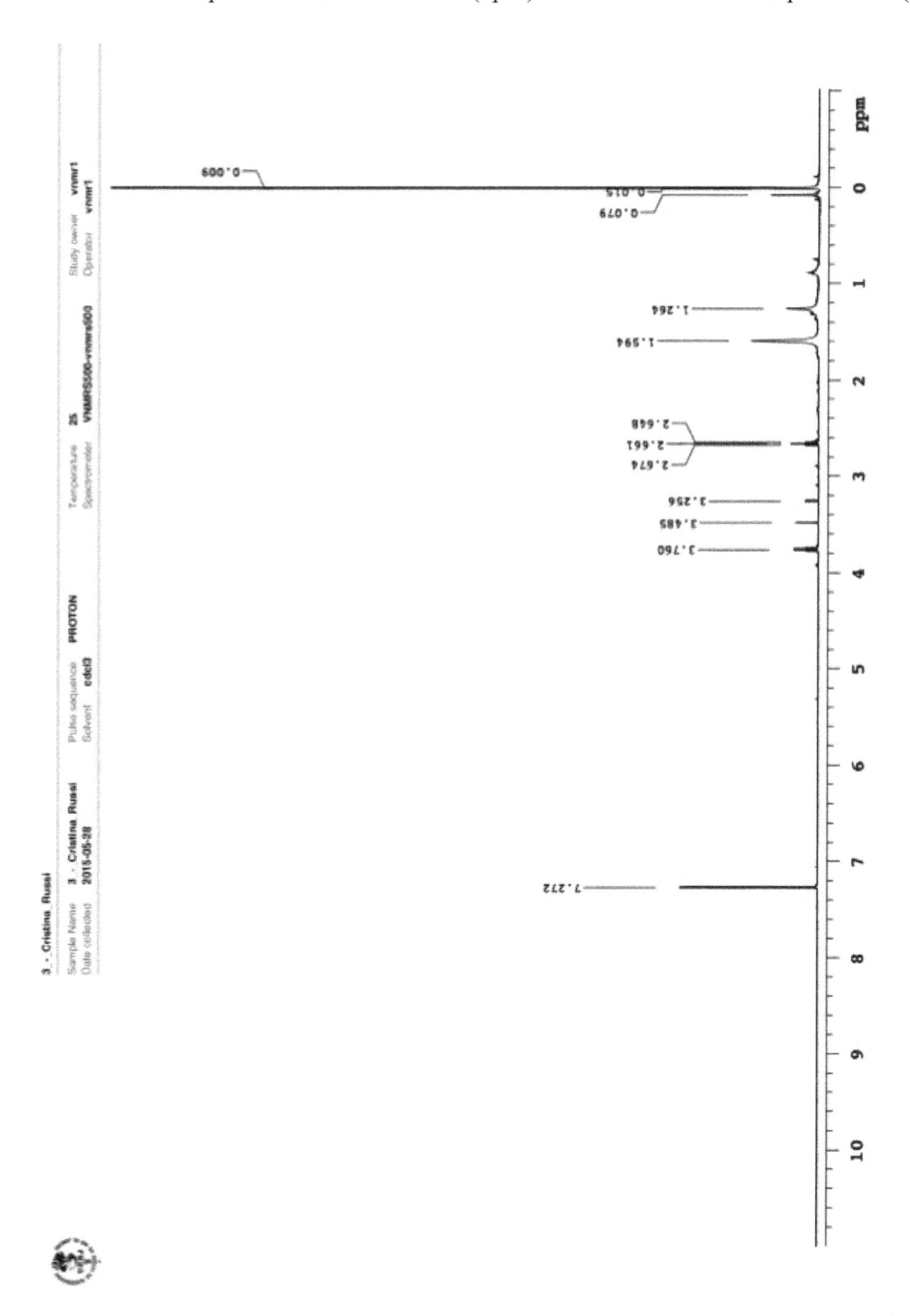

Figure 46 – H^{1}NMR spectrum of the solubilised surface components of the nitrile rubber composition with low TMTD (1phr) in the formulation. Composition: 3 (1/2/0.5).

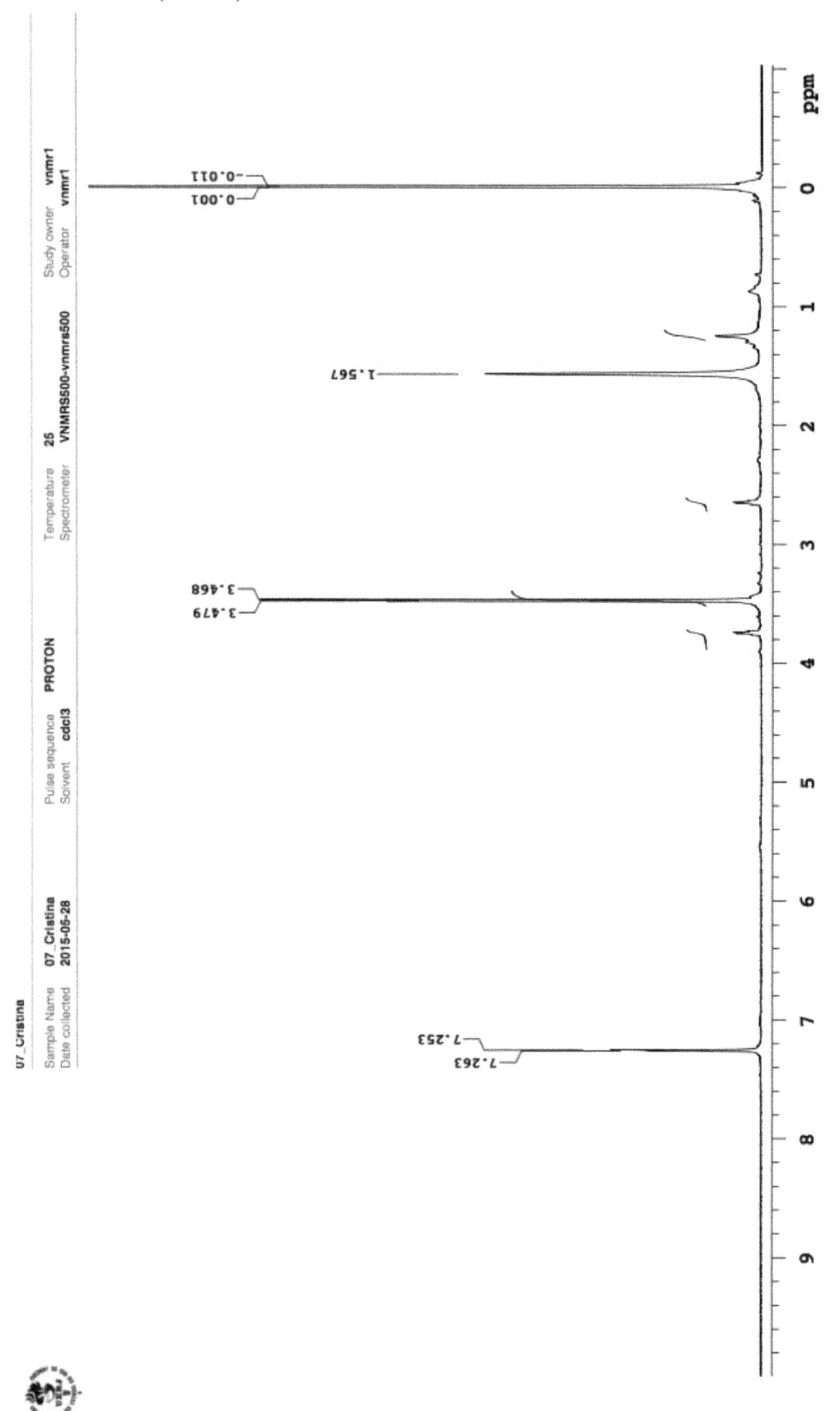

Figure 47 – H^{1}NMR spectrum of the solubilised surface components of the nitrile rubber composition with high TMTD (3 phr) content in the formulation. Composition: 7 (3/2/0.5).

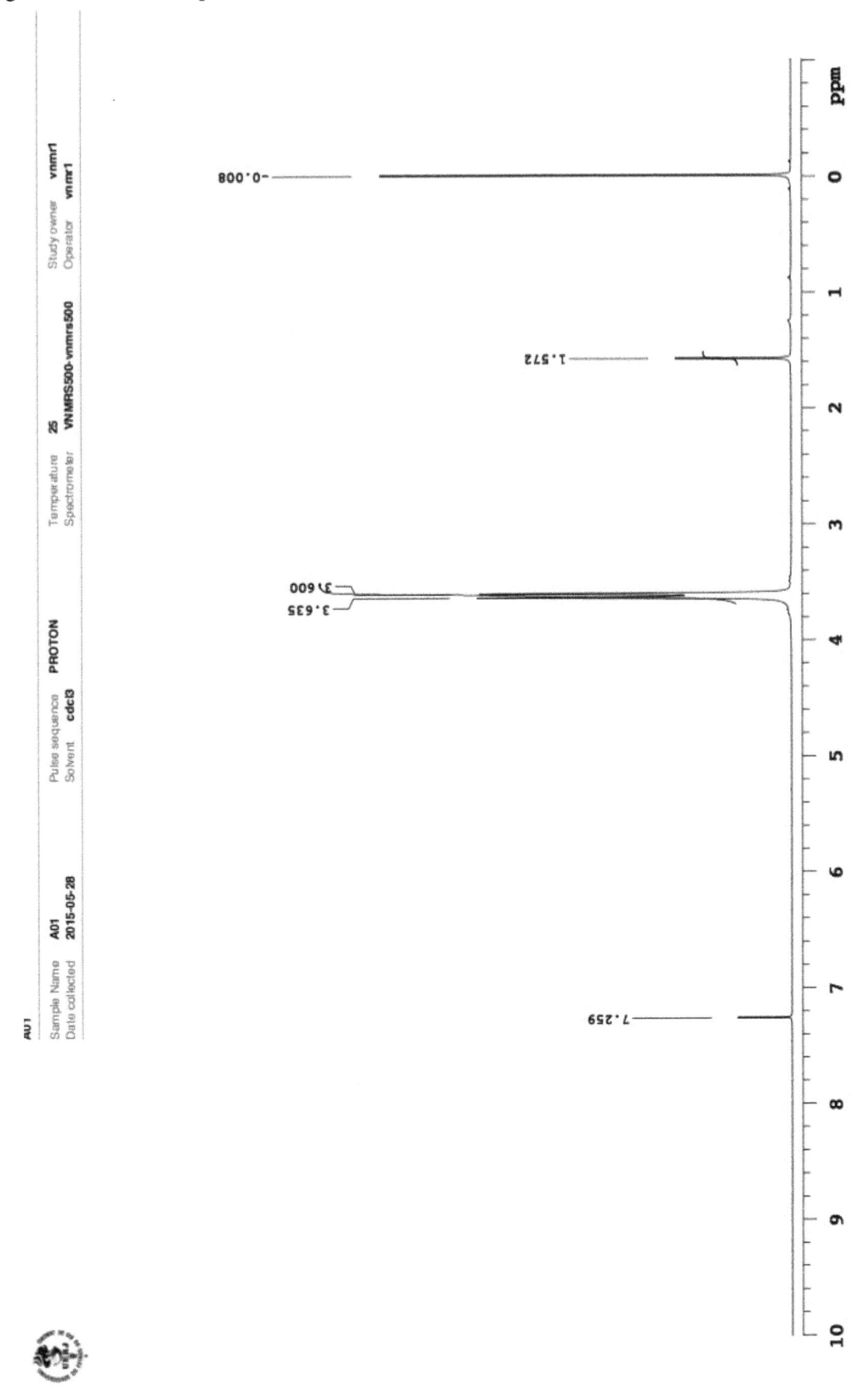

Figure 48 – H^{1}NMR spectrum of the accelerator TMTD.

Figure 49 – H^{1}NMR spectrum of the accelerator CBS.

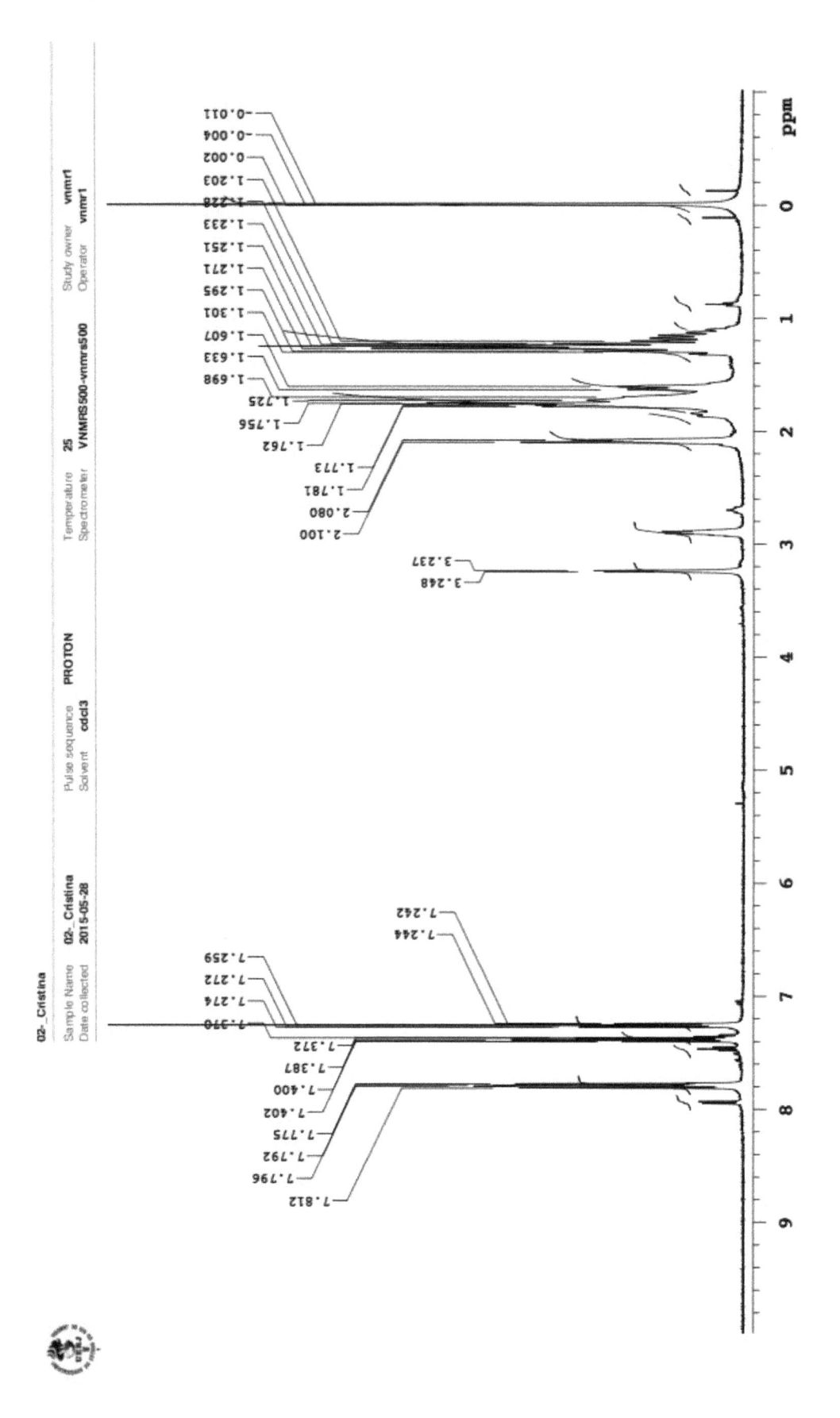

APPENDIX E

Figure 50 – Pareto chart of the stardardised effects variable on the mechanical properties. (a)Tensile Strength, (b)Elongation at break, and (c)Hardness.

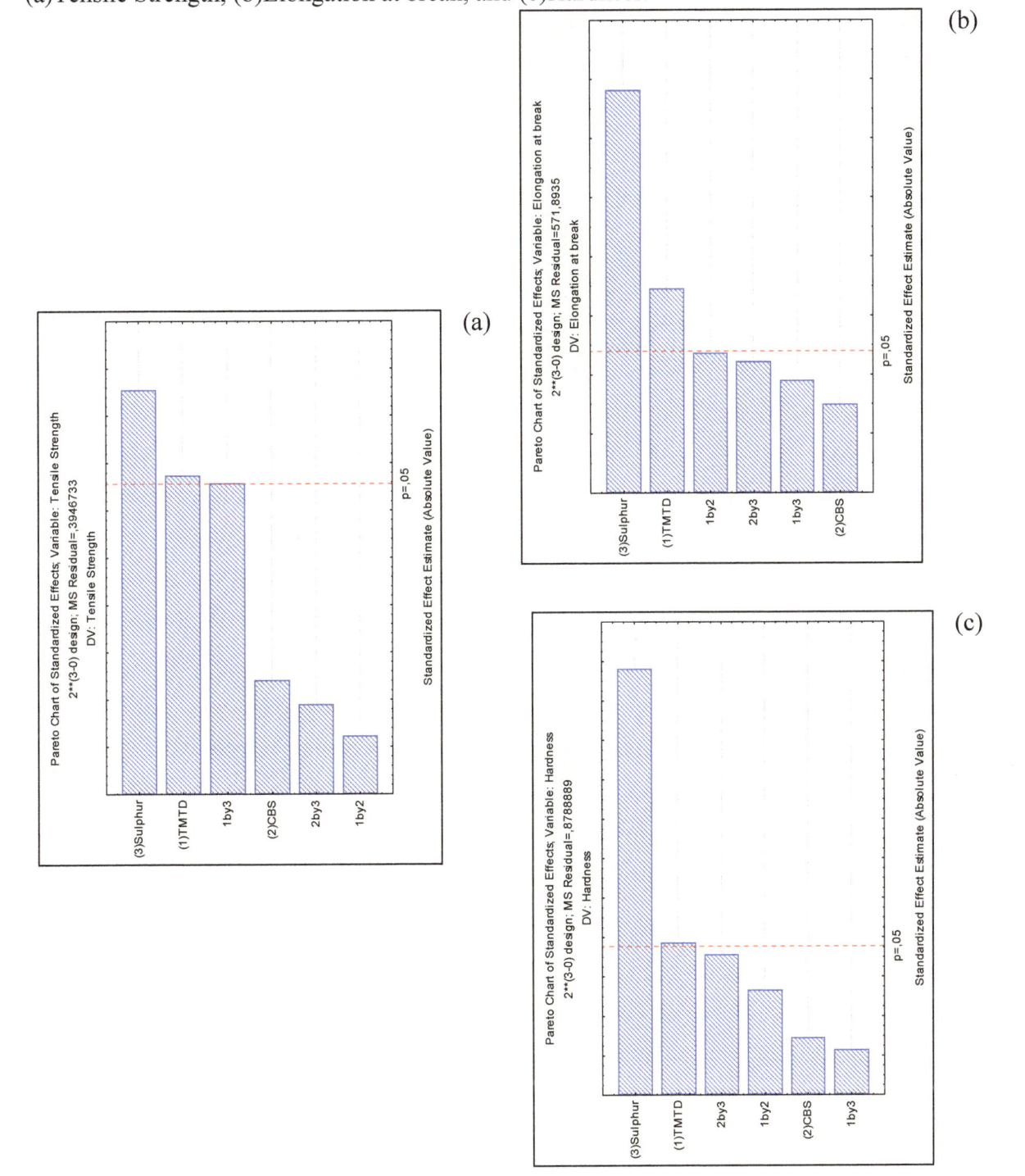

Figure 51 – Response surfaces for mechanical properties. (a) Tensile Strength; (b)Elongation at break; (c)Hardness.

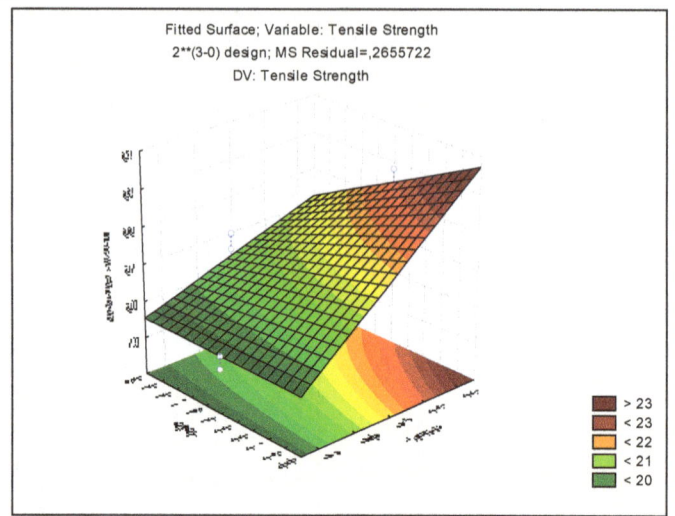

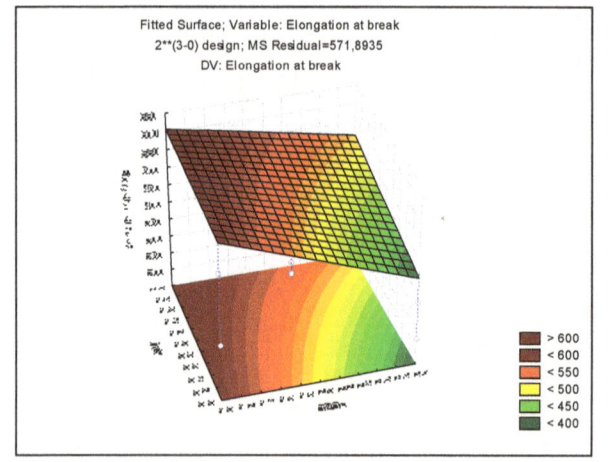

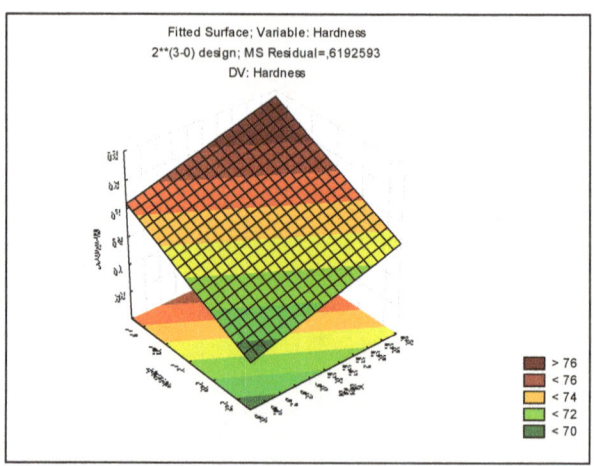

Figure 52 – Pareto chart of the stardardised effects variable on the change in mass after 22h of immersion in soybean biodiesel at 100°C.

Pareto Chart of Standardized Effects; Variable: Mass change 22h
2**(3-0) design; MS Residual=,0000478
DV: Mass change 22h

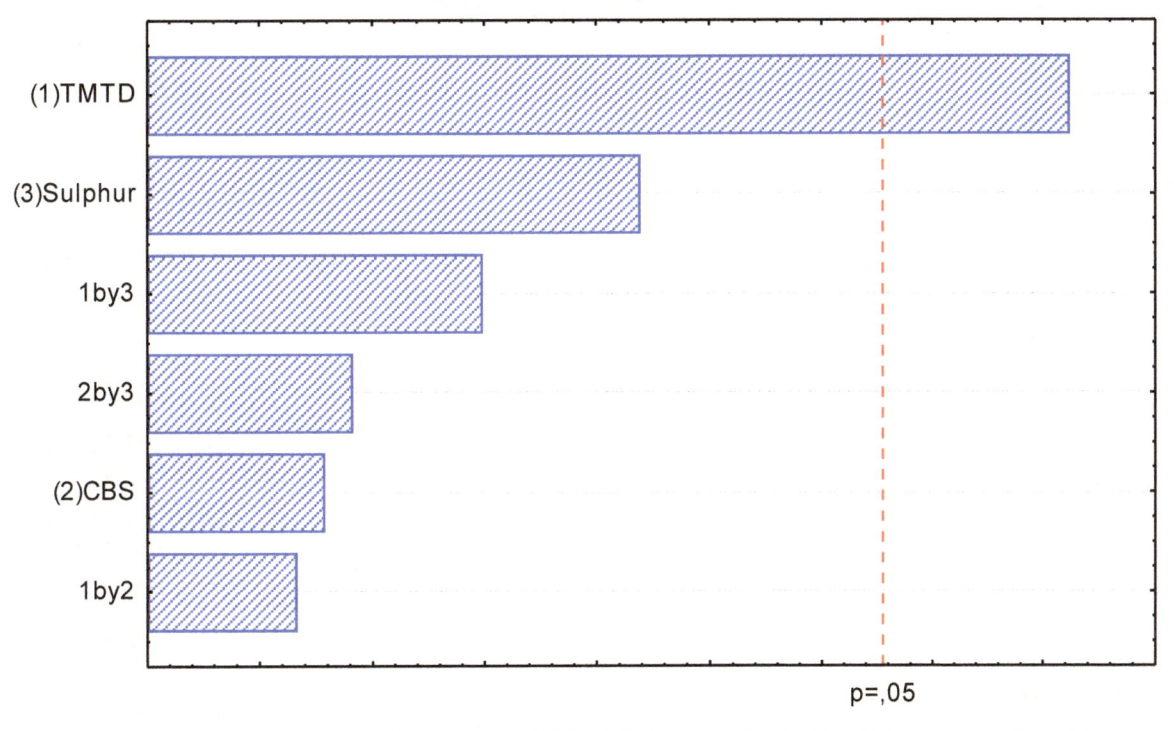

Standardized Effect Estimate (Absolute Value)

ANNEXE A

Table 16 - Chemical and physical properties of soybean biodiesel (Soybean methyl ester).

Property	Value	Unit	Method
Ester content	>99	% (m/m)	DIN EN 14103
Density (15°C)	884.9	kg/m^3	DIN EN ISO 12185
Kinematic viscosity (40°C)	4.118	mm^2/s	DIN EN ISO 3104
Flash point	166.0	°C	DIN EN ISO 3679
Cold filter plugging point (CFPP)	-7	°C	DIN EN 116
Sulphur	0.75	mg/kg	DIN EN ISO 20884
Carbon residue	0.20	% (w/w)	DIN EN ISO 10370
Cetane number	51.3	-	DIN EN 15195
Sulphated ash	<0.01	% (w/w)	ISO 3987
Water content	236	mg/kg	DIN EN ISO 12937
Total contamination	14	mg/kg	DIN EN 12662
Copper corrosion	1	Korr.-Grad	DIN EN ISO 2160
Oxidation stability	4.5	h	DIN EN 14112
Acid number	0.176	mg KOH/g oil	DIN EN 14104
Iodine value	130	g I$_2$/100g oil	DIN EN 14111
Linolenic acid methylester	7.4	% (w/w)	DIN EN 14103
Methyl ester (≥ 4 double bonds)	0.13	% (w/w)	DIN EN 15779
Methanol content	0.03	% (w/w)	DIN EN 14110
Free glycerine	0.01	% (w/w)	
Monoglycerides content	0.41	% (w/w)	
Diglycerides content	0.08	% (w/w)	DIN EN 14105
Triglycerides content	<0.01	% (w/w)	
Total glycerine	0.13	% (w/w)	
Phosphorous content	<0.5	mg/kg	DIN EN 14107
Sodium and potassium (combined)	1.8	mg/kg	DIN EN 14538
Calcium and magnesium (combined)	<0.5	mg/kg	

Source: Informed by the supplier (ASG Analytik-Service Gesellschaft mbH).

CPSIA information can be obtained
at www.ICGtesting.com
Printed in the USA
LVHW06s1034121018
593287LV00004B/18/P